The Amphibians *of*
Great Smoky Mountains
National Park

The Amphibians *of* Great Smoky Mountains National Park

C. Kenneth Dodd Jr.

Illustrations by Jacqualine Grant

THE UNIVERSITY OF TENNESSEE PRESS / Knoxville

All photographs taken by author unless otherwise noted. All animals were photographed within Great Smoky Mountains National Park unless otherwise noted. The color illustrations are by Jacqualine Grant.

This book is printed on acid-free paper.

Library of Congress Cataloging-in-Publication Data

Dodd, C. Kenneth.
The amphibians of Great Smoky Mountains National Park /
C. Kenneth Dodd Jr.; illustrations by Jacqualine Grant.
        p. cm.
Includes bibliographical references (p.   ).
ISBN 1-57233-275-1 (pbk.: alk. paper)
1. Amphibians—Great Smoky Mountains National Park (N.C. and Tenn.)
I. Title.
QL653.G76D64 2004
597.8'09768'89—dc22   2003018796

To the naturalists and scientists who have contributed to our knowledge and understanding of the amphibians of the Great Smokies and the southern mountains, and to those persons and organizations who protect the fauna and flora of these mountains today.

Most importantly, however, I dedicate this book to my wife, Marian Griffey, for all the hikes in the snow, rain, heat, and cold, for all the rocks and logs she turned, for the climbs up Mount Sterling and Siler's Bald, for her patience and dedication, but mostly for her friendship and her love of nature, and me.

Contents

Amphibians and the Environment in the Great Smoky Mountains National Park

Accounts of Species

Salamanders

Frogs

American Bullfrog *(Rana catesbeiana)* / 236
Northern Green Frog *(Rana clamitans)* / 239
Pickerel Frog *(Rana palustris)* / 243
Northern Leopard Frog *(Rana pipiens)* / 247
Wood Frog *(Rana sylvatica)* / 251
Eastern Spadefoot *(Scaphiopus holbrooki)* / 256

Preface

I first visited the Great Smoky Mountains in 1971. Having just returned from a year living in the tan, dry Sonoran Desert, I needed to see and smell green mountains again. I packed my stump ripper (a golf-club type of device with a hook used for turning over logs), camera, and notebook, and drove from my home in South Carolina to Indian Gap along the New-found Gap–Clingmans Dome road. There, I found my first red-cheeked Jordan's Salamanders, tiny Pigmy Salamanders, and a host of other multicolored, wriggling, and altogether wonderful beasts. During the course of my years at Clemson University, I often returned to the high ridges to take data on salamander behavior—or was it really just an excuse to watch the clouds roll through the spruce-fir forest on a crisp fall or spring day, and to smell the damp, coniferous earth? Of course I was not the first herpetologist to be drawn to these mountains, particularly to Indian Gap. It was here, after all, that the Imitator and Jordan's salamanders had first been discovered. The Pigmy Salamander was described from nearby Mount Le Conte. This was, and is, sacred ground to salamander enthusiasts.

After completing graduate school, my research took me away from the Smokies. In 1997, however, I received a call from Norita Chaney of the U.S. Geological Survey's (USGS) Inventory and Monitoring (I&M) Program, informing me that she had secured funding for an amphibian I&M pilot project in the southeastern United States. Would I be interested in heading the project, to be conducted at a National Park of my choosing? I thought of the Everglades in summer, and quickly decided that the Smokies would be ideal. As I drove from Sugarlands to Cades Cove during a fall day in 1997, and after realizing how big and inaccessible the Smokies were in many areas, I had my first (but not last) misgivings.

Over the course of the next four years, I assembled a group composed of enthusiastic young biologists and my recent bride to hike the hills and valleys, and to collect information on the species richness and distribution of the Park's amphibians. We used a variety of sampling techniques, including 10-×-10-meter survey plots, "permanent" 30-×-40-meter plots, litter bag surveys, and a great number of time-constrained litter and stream searches. We looked for previously reported rare species, sampled historic locations, investigated unique habitats, such as caves, and combed museum records and the published literature. Our efforts were designed to optimize the use of our available personnel within budget and logistic constraints, for even in four years with the most dedicated staff, some things could not get done. We examined more than 500 sites and recorded data on more than 10,000 amphibians. We were fortunate to walk in all parts of the park, in all seasons, and in all weather conditions.

The results from our quantitative research eventually will be published in the specialized scientific literature. However, our USGS crews turned up much new information on the natural history and distribution of the amphibians of the Great Smoky Mountains National Park—information that would be useful to park managers, other scientists, and to naturalists and visitors to the park. The present book is the result of my desire to present this information in one readily accessible source, much as James Huheey and Arthur Stupka had done nearly 30 years ago (Huheey and Stupka, 1967). *The Amphibians of Great Smoky Mountains National Park* provides new information, not just a popular summary of existing data. I hope that it will serve as a model for ongoing investigations of the biological diversity of public lands. Readers should keep in mind that the fieldwork was collaborative and that no one author can truly take credit for publishing the results of such a large-scale study (see acknowledgments). In that regard, I have tried to ensure that original observations are credited to the person or persons who made them. However, writing the book was my responsibility, and I have only myself to blame for errors in the text.

The purpose of this book is, therefore, twofold. First, I hope that the information contained in the book will be useful to the advanced naturalist and visitor, increasing his or her appreciation of the amphibians within the Great Smoky Mountains. Readers should keep in mind the three keys to enjoying natural history: 1) know the life history of the species in question; 2) understand the geology, environment, and other biota that share the land with the species or community of interest; and 3) examine past and present human influences on the species's distribution and biology. A naturalist cannot fully appreciate an animal or plant until these three areas are understood. The amphibian enthusiast must be part biologist, part landscape ecologist, and part historian. All of these things are connected.

Second, I hope to summarize the existing knowledge on the biology and distribution of the park's amphibians, both for professional biologists and park managers. The information presented—or, more importantly, the information as yet unknown—should spur additional observations and research. No one really likes to say "I don't know" when asked a question in his or her field of expertise, but there are far too many "I don't knows" out there with regard to our current understanding of amphibian biology, both within the park and elsewhere.

The reasons to go into the mountains, or indeed into any natural area apart from humans, extend far beyond administrative directives to conduct research on a particular problem. Hiking trails up and down mountains provides quiet time for contemplation and appreciation apart from a world seemingly obsessed with money, technology, and business practices. I may forget the computer programs used to store and analyze data, the hypothesis of interest at the time, or the details of the results. But I hope I never forget hiking up Rainbow Falls Trail in a driving rain through the misty, towering hemlocks; working in Gum Swamp on a beautiful autumn day as a mom bear and cubs strolled through the dry basin; seeing the huge Chinquapin Oaks above Laurel Falls, and the bear there lounging in a high tree; looking for salamanders among the wild flowers in Whiteoak Sink in early spring; watching hundreds of Marbled Salamanders at dusk on a rainy night; following Marian down countless trails and admiring the view and her long hair shining in the sun; or hearing the Spring Peepers, Upland Chorus Frogs, and American Toads calling from Cades Cove as a full moon lightened the mist rising from the twilight valley. As Emmett Reid Dunn said, "It is the places that stick in my mind." ∎

Acknowledgments

Assembling the data for this book has been a collaborative effort in all phases, particularly during the four years of hard field work between 1998 and 2001. In that regard, I thank the following individuals for their support, assistance, dedication, and companionship.

For providing the beautiful illustrations of larval amphibians: Jacqualine Grant.

For providing additional color photographs: Richard Bartlett.

For their field collecting skill, expertise, and general good humor during all kinds of weather conditions, the USGS field crew from 1998 to 2001 (in alphabetical order): Jamie Barichivich, Jeff Corser, Elizabeth Domingue, Marian Griffey, Kelly Irwin, Christy Morgan, Rick "Bubba" Owen, Kevin G. Smith, and Jayme Waldron.

For preparing the distribution maps: Alisa Coffin and Ann Foster.

For their support and assistance: the National Park Service personnel at Great Smoky Mountains National Park, particularly Keith Langdon, Dana Soehn, and Jack Piepenbring.

For library assistance and access to archival photographs: Annette Hartigan.

For access to the preserved amphibian research collection at Great Smoky Mountains National Park: Don DeFoe.

For assisting with housing: Jody Fleming (Discover Life in America).

For sharing information on the amphibians of the Great Smokies: Ben Cash, William Gutzke, James Huheey, James Petranka, Charles K. Smith, Stephen Tilley.

For providing funding: Norita Chaney (USGS Inventory and Monitoring Program; Reston, Virginia) and the USGS Amphibian Research and Monitoring Initiative (ARMI).

For providing research permits: National Park Service (Great Smoky Mountains National Park), North Carolina Wildlife Resources Commission, Tennessee Wildlife Resources Agency.

For reading the manuscript and making valuable comments: Marian Griffey, Dick Franz, Steve Johnson, Richard Seigel.

For editorial comments and advice: Scot Danforth.

For compiling the index and composing the poem "Acquainting": Marian Griffey.

Acquainting

I have learned something of myself on this day in the woods;
things no other teacher could have revealed.
Walking the steep path along a slick-rock stream,
my delight in nature's beauty reawakened,
as if for the first and countless time.
Wild iris cloaked a bank,
holding that patch of Earth together against the eroding ravages of
weather,
cupping their hoary throats into the wet heavens,
feeding human hungers with the meaning of life;
doing, without the anchor weight of "why?"
yet godly clothed in raiments beyond compare or price.
Solomon's Seal; Flea Bane; Dog Hobble; Wood Violet.
None coveting their neighbor's plot.
Then, down into the White Oak Sink,
where once a saw mill sang against the river's flow
and men shaped their living around the death
of Mother's tallest and oldest trees.
I nearly rolled down the grade, as, no doubt,
those ancient giants had done before me,
landing in the basin bruised but ready for my own new shaping
and the fashioning of a different look in the appearance of my living.
Hillsides adorned themselves in thick braids of Shooting Star, May Apple;
White and Nodding Trillium threaded trails of Columbine,
and filled chalices of fungal growth.

Pleasure grounds carried bouquets of Miniature Orchids,
and everywhere the tender salamanders haunted the disappearing
footprints of imagined
dinosaurs.
I touched, smelled, held, breathed the flavors
of yesterday and tomorrow . . . all within a single breath,
and knew it to be mine—this cosmos,
this postage stamp universe;
mine for the loving and mine for the nurturing.
Nature alone is just such a lover . . .
nurturing the nurturer;
giving satisfaction in the act of satisfaction taken
(when there is balance and no greed).
I crawled the up-most trail back to the beginning,
finding I had conquered, not the demands and limitations of a laurel-ized
mountain cove,
but rather my own assigned dreads and fears born of too many
fractioned
minds and too little symbiosis.
I learned something of myself in the woods today,
and walked out pleased for having made the acquaintance.

—Marian Lovene Griffey

Amphibians *and the*
Environment in the
Great Smoky Mountains
National Park

A Fascination for Naturalists

An Introduction to the Amphibians of the Great Smoky Mountains

Among all the little creatures of the woods and streams none have been so neglected as the salamanders and none are of greater interest. . . .

—Sherman C. Bishop, date unknown,
cited by Hunsinger (2001:241)

Although the eminent biologist Sherman Bishop only mentioned salamanders when he began his lecture notes in herpetology at Cornell
University, he could have easily substituted "amphibians" and still have
been accurate for his day. In the 1920s and 1930s, most people knew that
frogs called in the spring, and some knew, perhaps, that tadpoles turned
into frogs. In general, though, most understood little about the natural history of these remarkable animals. Today, many people have a much better
understanding and appreciation for the natural world of amphibians, in
part due to expanded educational opportunities in biology and environmental science, and in part due to news stories about the decline of many
amphibians. People who visit national parks are often keenly aware of nature
and come to the parks to learn more about the natural world. Visitors to
the Great Smokies nowadays are still liable to know more about frogs than
about the mysterious ways of salamanders, however.

I hope that through this book, naturalists, biologists, park visitors,
and armchair lovers of nature everywhere will come to have a greater

appreciation for the beauty, diversity, and complexity of the park's amphibian community. Appreciation comes from understanding and empathy comes from appreciation. In this first chapter, I discuss amphibian nomenclature, natural history, the kinds and diversity of amphibians, the complex life cycles of these species, and I give a brief history of amphibian research within the park. This and the following chapters will serve as a background for the species accounts of the more than 40 amphibians that inhabit the Great Smoky Mountains National Park.

Amphibian Names

People use three different types of names when talking about amphibians, and all describe the animal to one degree or another. Colloquial names are names used by people within specific areas of the country for an individual species or group of amphibians, but these have little or no recognition beyond the area of immediate use. In the South, the term "spring lizard," for example, is often used to refer to salamanders that are used as bait, regardless of the species involved, or whether the "lizard" is not a reptile at all, but an amphibian. Other colloquial names include waterdog, puppy dog, hop toad, and tree toad. In herpetology (the study of amphibians and reptiles), reptiles usually receive the more colorful colloquial names.

Common names are used by the general public and biologists alike to identify species. Some common names are often used incorrectly, sometimes to differentiate frogs from toads (all toads are in fact frogs) or salamanders from newts (both are salamanders). Common names often are descriptive of the animal, such as "Spotted Salamander" for the salamander *Ambystoma maculatum.* This species has large, bright orange, yellow, or red spots on its dark back, thus making the common name very appropriate. Sometimes colloquial names, such as mudpuppy, can even become accepted common names. In order to standardize common names as much as possible, herpetologists have recently published a common names guide (Crother 2000) for North American amphibians and reptiles. In the chapters and species accounts that follow, I use these recommended common names. In addition, I capitalize the common names, another recently recommended practice which heretofore was frowned upon by those who studied amphibians.

Throughout the world, there are many types of animals that could be given similar descriptive common names, such as "Spotted Salamander." In order to eliminate confusion, scientists use a third type of name, the scientific name, to describe a species unambiguously. No two species in the world may have the same scientific name. Scientific names are derived from Latin or Greek roots and have two or three parts. The first name is the genus (for example, *Ambystoma*), and the second name is the specific epithet (for example, *maculatum*). Thus, the name *Ambystoma maculatum* can only be used for the Spotted Salamander that is found in eastern North America. Genus names are capitalized, whereas specific epithets are not; both names are italicized or underlined. Placing two species within the same genus (for example, *Ambystoma opacum* and *A. talpoideum*) implies they had a shared evolutionary lineage at some distant point in their history.

As with common names, scientific names often, but not always, are descriptive of the animal named. For example, the scientific name of the Northern Slimy Salamander, *Plethodon glutinosus,* is derived from *Plethodon* (meaning "many teeth") and *glutinosus* (meaning "slimy"). Hence, it is a slimy salamander with many teeth, an apt description, especially if anyone tries to pick one up. Although they do not bite, they have many small teeth located on the jaws and roof of the mouth, and they are incredibly slimy. As an aid to understanding scientific names, an etymology is included with the species accounts that follow.

Species and Higher Taxonomy

Biologists have long debated the definition of what constitutes the fundamental unit of natural history, the species. Whereas it is very easy to distinguish a Spotted Salamander from a Marbled Salamander, distinctions made on appearance alone are sometimes very deceptive. Some animals appear similar to one another but are very different in their ecological requirements, genetics, or distribution. Others are very different in appearance, but are closely related. For example, some male amphibians are very different in appearance from females of the same species, which led early naturalists to identify them as different species. For other species, the appearance of juveniles is very different from that of the adults, such as the terrestrial eft and aquatic adult of the Eastern Red-spotted Newt. Today, the definition of "species" is far from resolved (see the reviews by Frost and Hillis [1990] and Frost [2000] for example), and it may never be possible to define species in such a manner that the term applies identically to animals, plants, bacteria, and all other living organisms.

For the purposes of this book, I use essentially a biological species concept, that is, a species is a group of potentially interbreeding organisms that produce viable offspring and that share a common evolutionary lineage. Species differ from one another in their ecological roles and sometimes in their appearance, and they usually do not interbreed with other species successfully. Unfortunately, the salamanders of the Great Smoky Mountains violate virtually every definition that attempts to pigeonhole them into well-defined categories, such as species. Certain species hybridize in narrow zones of contact in some areas, but not in others. In other species groups, there appear to be "hybrid-swarms" that occur over relatively large areas, although in other areas the two parent species act as reproductively distinct species. There is a large amount of genetic variation in some complexes that may or may not coincide with phenotypic variation.

Differences in species concepts, coupled with advances in genetics and population biology, have resulted in a complex taxonomic history for the salamanders in the Great Smoky Mountains; frog taxonomy has been more stable in this region. Names used in earlier field guides (Huheey and Stupka 1967; Conant and Collins 1991) often do not agree with the names used more recently (Huheey and Tilley 2001; this guide). Likewise, even recent summaries, such as Petranka's (1998) monumental work on salamanders or scientific journal articles may use names differently from one another, depending on the author's conceptions and interpretations of species. Although rearranging names is confusing to the naturalist and scientist alike at times, it points out that biology is a vigorous science, ever changing and never static. Likewise, species are dynamic and changing, and the more herpetologists study amphibians, the better they understand the complexity of their hidden lives. The variation seen in salamanders of the Southern Appalachians highlights the fact that the southern mountains constitute a great outdoor laboratory of evolution.

Higher taxonomic categories in biology are undergoing fundamental reassessment as biologists attempt to standardize definitions and to emphasize evolutionary history within classification schemes. All higher

Table 1

Families and Genera of Amphibians Found within Great Smoky Mountains National Park	
Family	Genera
Salamanders	
Ambystomatidae	*Ambystoma*
Cryptobranchidae	*Cryptobranchus*
Plethodontidae	*Aneides, Desmognathus, Eurycea, Gyrinophilus, Hemidactylium, Plethodon, Pseudotriton*
Proteidae	*Necturus*
Salamandridae	*Notophthalmus*
Frogs	
Bufonidae	*Bufo*
Hylidae	*Acris, Hyla, Pseudacris*
Microhylidae	*Gastrophryne*
Pelobatidae	*Scaphiopus*
Ranidae	*Rana*

taxonomic categories are human-derived; that is, they do not form natural or biological entities. In the most general terms, amphibians have generally been classified as follows: Kingdom Animalia; Phylum Chordata; Subphylum Vertebrata; Class Amphibia; Order Caudata (salamanders) or Anura (frogs). Orders are further broken down into Families. The Families of the amphibians living in the Great Smoky Mountains National Park are listed in table 1.

Body Plan

All salamanders and frogs within the Great Smoky Mountains are tetrapods, that is, they have four limbs. The front legs normally have four toes on each foot; the rear legs normally have five toes on each foot. However, there are exceptions. In the Smokies, the Four-toed Salamander and the Common Mudpuppy both have only four toes on each hind foot.

The loss of the rear toes extends to the larvae of these species, and the lack of the fifth toe helps to identify their larvae when compared with other salamander larvae. Salamanders and frogs sometimes lose toes during their daily activities, but these regrow readily and often are nearly impossible to distinguish from original toes.

Adults

Salamander bodies are elongate with a distinct head, a trunk, and a tail. In contrast with frogs, only salamanders have tails as both larvae and adults. The indentations on the side of the bodies of certain salamanders are termed costal grooves; these correspond to the location of the ribs. Although the number of grooves can vary within a species, counting them can help differentiate species from one another. The tail of a salamander is a remarkable structure. Tails can be important in the social behavior of the animal; they serve as a fat-storage organ to help survive periods of food scarcity, such as winter or dry spells; they store energy, as fats, to provide yolk for egg clutches; they often contain noxious or poisonous skin glands; they serve as a grasping or balancing organ for certain species; and they can be shed (autotomized) to thwart predation attempts. Salamander tails even have a specialized morphology between the tail vertebrae to allow for quick loss without a loss of blood or excess muscle mass. The basal tail constriction of the Four-toed Salamander may be a special adaptation to facilitate tail loss in an emergency. Shed tails wiggle and thrash, helping to draw attention away from the salamander's body, and tails are capable of regeneration. Whereas a lost tail can make a salamander less "socially" acceptable and may inhibit reproduction for a time, autotomy allows the animal to escape predation.

It may be difficult to tell male from female salamanders, especially during the nonbreeding season. Males of some species (for example, *Plethodon*) develop visible "mental" glands on the underside of their chins. These glands secrete chemicals important in mating. Other male salamanders (for example, *Eurycea*) develop elongations from the nostril opening to the upper jaw termed "cirri." The cirri have a superficial resemblance to external fleshy fangs, and they aid the salamander in picking up chemical cues from the environment. Male mole salamanders *(Ambystoma)* develop swollen cloacas and have papilla-like structures that line the cloaca. Female cloacal linings have a more folded appearance. In male newts *(Notophthalmus)*, the aquatic males develop a smooth, dark-green, lustrous skin, an expanded vent, a large tail fin, and black keratin-like structures on the toe tips and inner surface of the thighs. These rugose structures presumably aid the male in enticing and holding on to a receptive female.

Frogs, of course, lack tails as adults. Their short, powerful bodies with long hind legs are made for hopping or jumping, sometimes in great leaps. Toads *(Bufo)* usually walk or hop, whereas the so-called "true" frogs *(Rana)* are expert leapers. The toes of frogs normally are long and thin, except within the chorus and treefrog family (Hylidae). In this group, the ends of the toes are expanded slightly in cricket and chorus frogs *(Acris* and *Pseudacris)* or greatly in treefrogs *(Hyla)*. The treefrogs use their expanded toe tips to climb trees and smooth surfaces; the expanded surface area of the toe pad exerts hydric friction toward the surface on which the frog walks, thus enabling it to hold on and climb. The more ground-dwelling chorus frogs do not need such expanded toe pads. Aquatic species usually have a membranous web between their hind toes, which facilitates swimming.

Frogs do not have external ears, although they are very attuned to sound. The opening to the middle ear is covered by a thin, membranous structure called a tympanum, which is located behind the eyes. The tympanums of American Bullfrogs, for instance, are rounded and large; they are conspicuous on either side of the head. The tympanums of some other species are less easily seen. Many frogs have "warts," bumps or ridges on the back (the dorsum) or dorso-laterally. These bumps and ridges usually contain mucous glands, and are important for moisture retention and defense. The parotoid glands of toads are kidney-bean-shaped structures located on the head behind the eyes. Their shape and size are important for the identification of the different species of toads, and more about them is provided in the species accounts of the toads.

Male frogs of many species can be distinguished from females by the presence of vocal pouches and a darkened coloration on their throats, at least during the breeding seasons. Males also develop enlarged and roughened thumbs during the reproductive season that are useful while amplexing females. Female frogs are often much larger than males, and do not produce the "warning croak" when picked up. Outside the breeding season, differentiating the sexes may be very difficult.

Larvae

Salamander larvae look very much like small salamanders, with the addition of the bushy gills that aid in respiration in an aquatic environment. The color of the gills in larval dusky salamanders *(Desmognathus)* is white, whereas the gills of all other aquatic salamanders are red. Common Mudpuppies retain their gills as adults, but all other Smoky Mountain salamanders lose their gills at metamorphosis. Gills are highly enriched with blood vessels that extract oxygen from the surrounding water.

Biologists can tell important features about the life history of salamander larvae simply by looking at them. Larvae that live in swift-flowing water are streamlined and have reduced tail fins, short legs, and short, nonbushy gills. Larvae that live in ponds or rather still water are stocky in their body proportions and have wide tail and sometimes body fins, long legs, and long, bushy gills.

Frog larvae (tadpoles) have rounded, bulbous bodies powered by a tail as long as or longer than the body. Tadpoles essentially are feeding machines, with a mouth apparatus geared toward continuous feeding and a very long digestive tract. The type of mouth structure and the location of the anus and spiracles (the one or two openings where water exits after being pumped through the mouth for respiration and feeding) are characters that help identify the species. As with salamander larvae, the general body plan is correlated with the habitat where the tadpole lives. In species in which eggs are deposited in shallow ditches and stream backwaters, the tadpoles are compressed and have streamlined bodies and narrow tail fins. In species that deposit their eggs in deep ponds, the tadpole's body is usually deep and squat, with large tail fins and relatively short tails. The increased surface area mechanically helps the tadpole to swim in the stagnant water.

Amphibian Evolution

The first vertebrates to leave the water did so during the Devonian period, some 350 to 370 million years ago. These animals were transitional between the lobe-finned fishes and the true Amphibia. Paleontologists have shown that the precursor to modern amphibians moved around on land, based on the fossilized trackways that these animals left in soft muds. Their bony fossils, furthermore, clearly demonstrate that they were not adapted solely for living in the water, although they may have had fishlike tails. Some of the transitional forms, such as the giant flat-headed *Ichthyostega,* had impressive rows of teeth and a vertebral structure designed for flexibility and mobility outside water. The first true amphibians, the Labyrinthodonts, appeared in the Carboniferous period and survived until the early Cretaceous period, a span of about 230 million years. The Labyrinthodonts gave rise to the modern Amphibia and to the reptiles and were present throughout the age of the dinosaurs. Another group of primitive amphibians, the Lepospondyls, became extinct in the early Permian period, and left no modern descendants.

Modern amphibians evolved at least by the mid-Mesozoic era. A froglike amphibian, *Triadobatrachus,* is known from the Triassic period (about 240 million years ago), but the earliest known true frog *(Prosalirus)* was found in Lower Jurassic period deposits in Arizona. These deposits date from 180 to 185 million years ago. The oldest known salamander *(Karaurus)* was found in Upper Jurassic period deposits from southern Kazakhstan, making it roughly 150 million years old. Thus, both primitive and more modern amphibians were present throughout the Mesozoic era. When dinosaurs ruled the world, frogs called from the swamps and salamanders hid under stones and in murky wetlands. The basic body plan of modern frogs and salamanders was set nearly 200 million years ago, although the earliest froglike amphibians likely walked more than they hopped or jumped. Further information on amphibian evolution is provided by Hofrichter (2000).

Although they are conservative in body form, modern amphibians have evolved complex and remarkable life history patterns. The variation in life history is well illustrated in the ways salamanders in the Great Smokies cope with their environment, although the greatest variation in frog life histories occurs in the tropics. Modern amphibians are not in any sense "primitive" animals; they are highly evolved to survive under the conditions in which they normally live. Their diversity and abundance suggest that they are an extremely successful group of animals, or at least they were prior to entering modern civilization's twenty-first century.

Amphibian Richness and Biogeography

Throughout the world, there are approximately 4,700 species of amphibians currently described: more than 440 are salamanders, 165 are caecilians, and the rest are frogs. More than 240 amphibian species are found in North America north of tropical Mexico. Caecilians are little-known, tropical, legless amphibians, and are not found north of southern Mexico. They are not closely related to any amphibian species within eastern North America, so nothing further will be said about them here (see Hofrichter 2000 for more details on caecilians).

The greatest number of amphibian species occurs in the tropics, particularly among the frogs of South America, Australia, south Asia, and Africa, including Madagascar. There are no frog families that have their centers of species richness within the southern mountains, and only a few genera *(Acris, Pseudacris)* even have their greatest diversity within the southeastern United States. In many respects, frogs seem to be peripheral lowland invaders of the high Appalachians. The region of temperate North America with the greatest species richness of frogs is the South Atlantic Coastal Plain.

The very large salamander family Plethodontidae (the lungless salamanders), on the other hand, has two centers of species richness, one of which is in the Southern Appalachian region; the other is in the highlands of southern Mexico to Panama. This family includes 67 percent of the world's salamanders and, in the Appalachians, contains species with a wide diversity of life histories. The remaining salamander families are relatively small, often containing only a handful of species. The newts (Salamandridae) are mostly Eurasian; the primitive hynobiids (Hynobiidae) occur only in Asia; the torrent (Rhyacotritonidae) and Pacific giant salamanders (Dicamptodontidae) are found in the Pacific Northwest; and two of the three species of giant salamanders (Cryptobranchidae) occur in temperate Asia. The remaining four salamander families have their greatest diversity in eastern, temperate North America, albeit not in the southern mountains. These families are the mole salamanders (Ambystomatidae), amphiumas (Amphiumidae), mudpuppies (Proteidae), and sirens (Sirenidae). The newt, cryptobranchid, mole salamander, and mudpuppy families are all represented in the Smokies. A much more detailed analysis of the biogeography of North American amphibians is given by Duellman and Sweet (1999).

Species Richness in the Smokies

There are 31 species of salamanders and 13 species of frogs that have been reported historically from the Great Smoky Mountains National Park. The Green Salamander may be extirpated from the park and likely never was abundant. The Northern Cricket Frog was reported from the park based on records from Chilhowee, Tennessee (Huheey and Stupka 1967), but probably never occurred within park boundaries. The Northern Leopard Frog may be extirpated, although there was a possible sighting in 2000. More historical information on these species is found in their respective species accounts.

The remaining 40 species are not distributed evenly throughout the Great Smokies. Some salamanders normally are found only in the higher elevations (for example, Jordan's Salamander, Southern Gray-cheeked Salamander, Ocoee Salamander), although they are not, perhaps, as exclusively confined to these habitats as previously believed. Others (the *Ambystoma,* the Four-toed Salamander, Southern Red-backed Salamander) are lowland species. A few salamanders (the Black-bellied Salamander, Spring Salamander) span a great range of elevations. Likewise, a few salamanders are found only in Tennessee (the *Ambystoma*), whereas others essentially are found only in North Carolina (Three-lined Salamander, Seepage Salamander). Some salamanders have extremely restricted distributions within the park, reflecting specialized habitat requirements (for example, Cave Salamander, Southern Zigzag Salamander), whereas others are widespread and occur in many types of habitats (for example, the Black-bellied Salamander).

Undoubtedly, current amphibian distribution patterns are the outcome of a complex interaction of evolutionary divergence, historical biogeography, and specialized habitat requirements, even without considering the influence of humans. The interplay between evolution, geography, and ecology has resulted in a high species richness of salamanders within the park, but the same is not true of frogs. Frogs require aquatic breeding sites, and these are scarce in the mountains, especially on the south and southwestern sides of the park. Hence, few species of frogs are found in the Smokies relative to the Atlantic Coastal Plain or the Tennessee Valley.

There are three nonhuman-related historical and biological factors that contribute to the high species richness of salamanders within the

Great Smokies and to the patterns of amphibian distribution in general within this national park. These are:

1. *The Southern Appalachians are the center of evolutionary divergence for the lungless salamanders, family Plethodontidae.* Because of the great diversity in life histories and the large number of species, the family Plethodontidae has long been thought to have evolved in the region of what is now the Southern Appalachians, including the Great Smokies. Plethodontid and ambystomatid salamanders were hypothesized to be closely related, with a lunged ancestor giving rise to the lungless descendant and to the mole salamanders. New molecular evidence has caused herpetologists to alter their ideas of salamander evolution, however.

Early hypotheses also suggested that the lungless salamanders lost their lungs as an adaptation to life in the cold, well-oxygenated mountain streams of the early Appalachian highlands. In such habitats, lungs would be disadvantageous, since the buoyancy of the lungs would cause the animals, especially larvae, to be swept downstream, and the cold, well-oxygenated water would negate the necessity of lungs, since oxygen diffuses readily across the skin surfaces of amphibians in cold water. New geologic data have caused biologists to question this explanation as well, although the region of the southern mountains is still thought to be the center of evolutionary divergence for this particular family.

Geology and molecular biology have changed scientific explanations of how the Plethodontidae evolved. During the late Cretaceous period to early Cenozoic era, geologists now know that the Southern Appalachians were not the mountains of today. Instead of craggy or rounded peaks with cold, rushing streams, the area was a vast, subtropical, lowland plain with intermittent rolling topography, much like the Gulf Coastal Plain of Alabama, Georgia, and panhandle Florida of today. Because there were no mountains and the climate was not the same as today, the lungless salamanders could not have evolved in cold mountain streams, and lunglessness could not have evolved as an adaptation to such habitats.

Current hypotheses suggest that the early ancestor of plethodontids gave rise to two lineages in the region of the present Appalachians. One lineage became the amphiumas, which are still found in southeastern swampy lowlands, whereas the other evolved into the modern plethodontids. The former specialized in muddy, mucky habitats, whereas the proto-

Predation and Further Defense

Despite their secretions, amphibians are eaten by a wide array of predators. Mammals, including raccoons, opossums, bears, otters, coyotes, wild pigs, and shrews, certainly eat amphibians whenever they are encountered. Raccoons even learn to eviscerate toads and frogs with noxious secretions from the "bottom side up," since the granular glands are located on the frog's back and sides. Birds forage through the leaf litter and attack salamanders there. During our sampling in the Smokies, we commonly found injured salamanders that looked as if they had been attacked by birds. In aquatic environments, amphibian larvae are consumed by many invertebrates, especially predaceous diving beetles and dragonfly larvae. Amphibians are eaten by reptiles, especially water snakes and small terrestrial snakes, such as garter *(Thamnophis)* and ring-necked *(Diadophis)* snakes. And, of course, they are eaten by other amphibians. The huge biomass of amphibians undoubtedly fuels the energetics of any ecosystem where they are present, at least during the reproductive and larval seasons.

In addition to secretions, amphibians have evolved coloration and pattern defenses that allow them to blend into their environment (camouflage) or to advertise their noxious or poisonous qualities (termed "aposematic coloration"). The lichenlike green and brown coloration of the Green Salamander makes it very difficult to see, as do the brown shades and patterns of the dusky salamanders *(Desmognathus)*. On the other hand, the bright orange coloration of the eft phase of the Eastern Red-spotted Newt, or the striking spot patterns or cheek patches of some other salamanders (for example, Jordan's Salamander, the *Ambystoma*), advertise that they are best left alone. A few salamanders that are edible have evolved color patterns similar to those that are noxious, thus rendering them somewhat less likely to be attacked. This type of defense (termed "Batesian mimicry") is found among several salamander groups in the Great Smoky Mountains.

A Chemical World

Salamanders are very attuned to the chemical world about which biologists are only beginning to discover. Woodland salamanders *(Plethodon)* mark territories with chemicals that convey messages to conspecific males and females. For example, a female can assess the quality of the territory inhabited by a resident male by sniffing the feces of the male. If he has eaten a high-quality food, such as termites, she is more likely to respond to him than if he has eaten a low-quality food, such as ants. Chemical marking allows salamanders to space themselves within the environment and to ensure that only the best animals maintain quality territories. Chemical cues (termed "pheromones") also play important roles in the courting rituals of salamanders by stimulating females to pick up sperm packets (termed "spermatophores") that are deposited on stalked jelly structures. Fertilization is internal for all species of Smokies salamanders, except the Hellbender. In this species, the female deposits her eggs in a nest constructed by the male; he then sheds his sperm directly onto the egg strings.

Mating and Reproduction

Whereas chemical communication is important in salamander behavior, it appears to be of less significance to frogs within the Smokies. Male frogs gather at breeding sites from winter to early summer and send out chirps, peeps, and trills, depending on species. The calls serve two main functions: to entice the females to the pond and to inform other males that the caller has occupied a particular portion of the pond. The female can assess male quality by the sound of his call, by how long it is, by how loud it is, or by how deep it is. Inasmuch as some frogs are territorial at breeding sites, the calls serve to alert other males that the caller is ready to defend his portion of the pond. Females often prefer the largest males when presented with a choice. On the other hand, "satellite" males (males that do not call but sit near a calling male) sometimes intercept females on their way to breed and thus avoid the competitive chorus.

In contrast to salamanders, all frogs in the Smokies have external fertilization. The process of holding onto a female during reproduction is termed "amplexus." The male frog grabs the female dorsally either under the armpits (axial amplexus) or just in front of her hind legs (inguinal amplexus) and holds on as strongly as he can. The location where the male grabs the female is species-specific. As eggs are extruded from the female's vent, he releases clouds of sperm over them. Male frogs often amplex the wrong species, other males, or inanimate objects. If another male is amplexed, he gives a warning croak and/or vibration to let the courting male know that he has erred in his mate choice. Otherwise, a male amplexing an inappropriate object often has a long and frustrating reproductive season.

Frogs and pond-breeding salamanders generally return to the same ponds or wetlands to breed from one year to the next, although there are exceptions. In some species of frogs, both sexes may use the same ponds each breeding season. In other species the males are site-specific, but the females are not. In this way, a female can choose the best fit male among all the males that she can hear, regardless of where he is located. Male pond-breeding salamanders and frogs usually arrive at the ponds first and establish their calling sites (for frogs) and territories. In Wood Frogs, the males overwinter closer to the ponds than females, presumably so they can get to the breeding sites early. Males also stay at ponds longer than females, who only stay long enough to mate and deposit their eggs clutches.

Activity

Amphibians are especially active at night. They congregate around breeding ponds, emerge from under rocks and logs, and forage through the terrestrial leaf litter or along stream edges. Amphibians often can be spotted easily with a flashlight as they sit at the mouths of their burrows and hiding places, waiting for unsuspecting invertebrates. At night, aquatic species and larvae also leave leaf litter mats and other hiding places, becoming conspicuous as they walk across stream and riverbottoms. Only on rainy nights, especially, can the abundance of amphibians be appreciated as they move over the forest floor, or even climb trees. Many amphibians are active by day, but they are more secretive, hiding under surface debris or wedging themselves under tree bark or in rock crevices. Presumably, nocturnal activity makes them less prone to predation, especially by largely visually oriented predators, such as birds.

Amphibian Life Cycles

Aquatic and Terrestrial Habitats

When most people think of amphibian life cycles, they think of the "typical" pattern whereby a salamander or frog moves to a wetland to breed, lays eggs, and then moves back to the terrestrial habitat to forage or overwinter.

 Small stream, The Sinks.

The eggs develop into larvae that remain in the wetland for a period of time, the larvae metamorphose, and the tiny juveniles disperse. In fact, only 73 of 142 or so amphibians in the southeastern United States follow this pattern. In the Great Smokies, all of the frogs and 19 of the salamander species have a life cycle that varies between wetland breeding sites and the surrounding terrestrial habitats. The wetland breeding sites vary between sinkhole and depression ponds (favored by *Ambystoma, Notophthalmus,* and many frogs), to seeps and small trickles (some *Eurycea* and *Desmognathus*), to fast-flowing streams and even good-sized rivers (Junaluska Salamander, Black-bellied Salamander, American Bullfrog).

Those amphibians that breed in small temporary or semi-permanent ponds do so because these habitats lack fish predators. Small seeps, especially at high elevations, also lack fish, but, like the larger streams, they are usually inhabited by other salamander predators. In areas lacking aquatic vertebrate predators, larvae are often conspicuous; in habitats where predators are abundant, larvae are secretive and cryptic. Many of the ponds and streams in the Smokies are temporary or intermittent. As the year progresses toward summer, these habitats dry up, making them unsuitable for amphibian breeding. Drought and low-rainfall years can also pose problems, and in some years amphibians may not be able to breed. Thus, whereas amphibians breeding in temporary wetlands may reduce predation risk to their larvae, there is a trade-off as to whether the wetlands will fill and allow sufficient time for the larvae to develop and metamorphose.

As discussed earlier, eggs of frogs and some species of salamanders are deposited in water and develop within it. The larval period will last a few days to as long as two or more years, depending on species. Most larvae transform within a year, however. If breeding occurs in spring or early summer, transformation occurs in late summer to autumn. If breeding occurs in the fall, the larvae overwinter and metamorphose the following summer. There is a great deal of variation in the larval period among species, however. Some of the variation undoubtedly is derived from hereditary factors, although the availability of high-quality food and the duration that water remains in the wetland (termed the "hydroperiod") also influence the length of the larval period. After transformation, the young juveniles disperse into terrestrial habitats.

Whereas many amphibians travel short distances from their breeding sites, others travel hundreds of meters or even several kilometers from where they passed the larval stage. Terrestrial habitats may be as close as streamsides, making certain salamanders (for example, the Black-bellied Salamander) truly semi-aquatic. Other salamanders (for example, the Black-chinned Red Salamander) and most frogs travel extensively before establishing a nonbreeding terrestrial home range. Here, they spend their time in surface, subsurface, or even in arboreal habitats for one or more years before returning to the area where, presumably, they hatched.

Some aquatic-breeding amphibians may spend years in terrestrial habitats before they begin their adult breeding cycles. Others may breed the first year after metamorphosis. As a rule, the shorter-lived an amphibian is, the sooner it begins to breed. Still, not every adult amphibian breeds every year, and little is known about reproductive periodicity through their life span. I have observed a female Florida Eastern Narrow-mouthed Toad skipping a breeding season, even though the radioactively tagged animal was near a pond and the life span of this species is only about four years. Perhaps females, in particular, skip breeding in years of harsh environmental conditions and poor food availability, and instead keep their energy reserves to stay alive.

Aquatic Habitats

In contrast to the "typical" cycle, two amphibians in the Great Smoky Mountains never leave the water, and instead carry out their entire life cycle in aquatic habitats. These salamanders are the Hellbender and the Common Mudpuppy. Both are active more at night than during the day, when they hide under rocks and other bottom debris. Little is known about the life histories of these species within the Great Smokies, but both lay eggs that hatch into larvae. The larvae are very difficult to find. Larval Common Mudpuppies have been found associated with leaf mats that form in the autumn in larger and slower-flowing creeks, such as Abrams Creek. Larval Hellbenders likely bury down into the interstitial gravels of the larger streams, and have been found in Little River. Hellbenders are territorial during the mating season, but probably not at other times.

Hellbenders, at least, may be exceptionally long-lived in their stable stream habitats. Specimens in captivity have been kept for more than 50 years. In pre-European times, these permanently aquatic species undoubtedly were more common than they are today because of recent habitat degradation, especially during the logging period. In the Ozarks, one study estimated that there were 468 Hellbenders in a single 4.6 km stretch of stream habitat, but nothing is known about historic or present abundance in the Great Smokies. Unfortunately, the Ozark population has since been decimated by habitat changes and collection for the pet trade. Nothing is known about the density of Common Mudpuppies in southern streams. Both Common Mudpuppies and Hellbenders usually remain within 100 meters or so of their favored stream sections, although Hellbenders, at least, have been documented moving up to one kilometer before returning. More details of the life histories of both of these species in the Smokies are found in the species accounts.

Terrestrial Habitats

There is another group of salamanders in the Great Smokies that never venture to water to breed: the entirely terrestrial plethodontid (lungless)

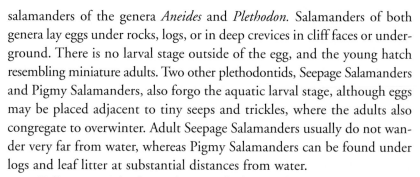

salamanders of the genera *Aneides* and *Plethodon*. Salamanders of both genera lay eggs under rocks, logs, or in deep crevices in cliff faces or underground. There is no larval stage outside of the egg, and the young hatch resembling miniature adults. Two other plethodontids, Seepage Salamanders and Pigmy Salamanders, also forgo the aquatic larval stage, although eggs may be placed adjacent to tiny seeps and trickles, where the adults also congregate to overwinter. Adult Seepage Salamanders usually do not wander very far from water, whereas Pigmy Salamanders can be found under logs and leaf litter at substantial distances from water.

Fully terrestrial salamanders usually have rather small home ranges on the surface of the ground, and some may defend territories from intruders, especially when food or shelter is scarce. However, their terrestrial homes include more than just the land surface, as these species may spend most of their lives below the surface in cracks, crevices, tunnels made by burrowing animals, or in old root channels. Some species, such as Pigmy Salamanders and Blue Ridge Two-lined Salamanders, leave the forest floor on rainy nights and climb tree trunks and limbs, presumably to forage for arboreal insects. Thus, amphibians using terrestrial habitats have both a horizontal and a vertical aspect to their habitat preferences. When looking for salamanders or trying to conduct studies on them, these multidimensional habitat requirements (surface, underground, arboreal) need to be taken into consideration.

Terrestrial salamander populations may attain rather remarkable numbers, and the biomass of salamanders in an area may exceed that of mammals, birds, and reptiles, making them very important in ecosystem dynamics. In a study of the salamanders of Walker Creek in Buncombe County, North Carolina, Petranka and Murray (2001) estimated that 18,486 salamanders, weighing 16.5 kg, were found on each hectare (2.2 acres) of old-growth mountain forest. These figures included 10,666 dusky salamanders *(Desmognathus)* and 2,600 woodland salamanders *(Plethodon)* of three species. During surface removal sampling, 672 *D. wrighti* were captured without any evidence that the population was impacted. Such a study emphasizes the three-dimensional habitat use of these relatively nonmobile animals. Biologists have much to learn about salamander numbers and distribution underground, and about how and why they move between surface and subsurface habitats.

Studies on the Amphibians of the Smokies

The Early Years

In order to understand the biology of amphibians in the Great Smokies and elsewhere in the Southern Appalachians, naturalists and scientists alike have been coming to the mountains for a long time. The names of early naturalists are memorialized today in Smokies place names, such as Mount Le Conte, named after Georgia naturalist John Le Conte, and Mount Guyot, named for geologist Arnold Guyot. The first naturalists and many park visitors undoubtedly recognized the wealth of salamander diversity they found, even if they did not appreciate its full extent.

One of the first recorded herpetologists to visit the Smokies was Emmett Reid Dunn, the father of the modern study of salamander biology. Dunn, first at Smith College and later at Haverford College, worked throughout the mountains of western North Carolina gathering data on species and their distribution and collected extensively in the vicinity of Mount Sterling in 1919. This area was very different than it is today, with a bustling town servicing the Suncrest Lumber Company on Big Creek. Indeed, much of the area had already been severely logged by the time Dunn arrived. The specimens that he collected eventually were deposited at the Museum of Comparative Zoology at Harvard, and included the Three-lined Salamander in an area where it has not been collected since. Dunn's work in the Southern Appalachians culminated in the 1926 publication of the monumental *Salamanders of the Family Plethodontidae* (reissued in 1972 by the Society for the Study of Amphibians and Reptiles).

Following Dunn's foray, the most significant collecting in the Smokies was carried out by Worth Hamilton Weller, an 18-year-old amateur biologist who was fascinated by salamanders. Weller visited the Smokies in 1929 and 1930 and collected salamanders in the Mount Le Conte vicinity, eventually describing a yellow-cheeked dusky salamander as *Desmognathus aureatagulus* (Weller 1930a). This species was later recognized as a color variant of the Imitator Salamander, described by Dunn in 1927. Weller also reported the presence of the Green Salamander within the Smokies (Weller 1930b) and authored the first complete checklist of the salamanders known from the park (Weller 1931). Just before he was to enter Haverford College in 1931 to study under Dunn, Weller died in a fall on Grandfather Mountain, where he had collected the first gold-backed specimens of a new species, subsequently named in his honor *(Plethodon welleri).*

The National Park Service

With the coming of Great Smoky Mountains National Park in 1934, two National Park Service employees, Willis King and Arthur Stupka, substantially increased knowledge of the herpetofauna of the Park. King was the park's first official wildlife biologist; he received his doctorate in 1939 based on a study of the amphibians and reptiles in the Great Smokies (King 1939a). He soon published this work (King 1939b) and later added further notes on amphibian and reptile biology within the park (King 1944). The Pigmy Salamander was first described by King based on specimens collected at Mount Le Conte (King 1936). As was common in those days, King's study combined systematics, morphology, ecology, and biogeography, and was the first study to synthesize what was known about herpetology in the Smokies. Early in the 1930s, King, as a wildlife technician, was involved in some controversial decisions. He instructed Park Service personnel in killing water snakes *(Nerodia)* to "improve" fishing, and he supported the introduction of non-indigenous rainbow trout. Neither practice would be acceptable today.

The first official National Park Service naturalist in the Great Smokies, Arthur Stupka, was also very influential in understanding the biology of the amphibians in Great Smoky Mountains National Park. As

Willis King, first wildlife biologist in Great Smoky Mountains National Park. Photograph taken in 1939. GSMNP file IV-P-17912.

Arthur Stupka, first park naturalist in Great Smoky Mountains National Park. Photograph taken in 1946 by S. G. Baldwin. GSMNP file IV-P-12946.

recounted by Brown (2000), Stupka cataloged the park's biota and recorded detailed natural history observations from 1935 to 1966, when he retired. He then teamed up with James Huheey of the University of Maryland to publish the first guide to the amphibians and reptiles of the Smokies, which was also the first true field guide devoted entirely to the herpetofauna of any national park (Huheey and Stupka 1967). The herpetology guide was more than a simple identification manual. Instead, it drew upon Stupka's 30 years of experience of recording detailed notes on the park's amphibians. He recounted where they occurred, when they reproduced, what their activities were. Stupka authored or co-authored four additional books on the animals and plants of the Great Smokies. Arthur Stupka died in April 1999 at the age of 93.

Although a chemistry professor, Huheey had become interested in the salamander genus *Desmognathus,* a group still considered difficult to classify at best. He published a paper on the dusky salamanders of the Smokies (Huheey 1966) and conducted studies on salamander warning coloration and mimicry. In 2001, recently retired, he and Stephen Tilley published a small guide to the herpetofauna of the Smokies for the Great Smoky Mountains Natural History Association (Tilley and Huheey 2001).

The Modern Research Era

The era of modern amphibian research in the Smokies began in the 1960s with the systematic studies by Richard Highton of the University of Maryland and of Stephen Tilley of Smith College, E. R. Dunn's first academic post. Both are indefatigable field biologists who began their careers with studies of salamander morphology and descriptive ecology. As their careers progressed, they employed the latest sophisticated techniques in molecular biology and quantitative population ecology to attempt to unravel the mysteries of salamander evolution in the Southern Appalachians. Thanks to Highton's studies on *Plethodon* and Tilley's work on *Desmognathus,* herpetologists are beginning to understand and appreciate what a marvel of biology a salamander is. Both continue their studies today in the Smokies, particularly on the phylogeny and evolution of mountain species (Tilley 1981, 1988, 2000; Highton 1989; Highton and Peabody 2000).

Along the Balsam Mountain Road on Balsam Mountain, Nelson Hairston Sr. of the University of North Carolina has conducted pioneering research on salamander ecology. Beginning in the early 1970s, Hairston and his students have monitored trends in the abundance of salamanders at specific locations (Hairston and Wiley 1993), and they have studied how competition structures salamander communities (Hairston 1987), specifically across hybridization contact zones (Hairston et al. 1992). Aggression is more evident in areas of strong competition, but less so in other areas. Hairston's studies demonstrate that competition is a driving force in the spacing and richness of salamander populations, and it acts differently among species and in different ecological situations.

Through the years, other amphibian specialists working in the Smokies have contributed greatly to understanding how nature works. Edmund Brodie Jr. (Utah State University) has conducted pioneering work on salamander defensive behavior; his research received an early start through his observations of Jordan's Salamander and its red-cheeked mimic in the Smokies. Richard Bruce (Western Carolina University) has delved into the evolution of salamander life history based on research, in part, on the primitive aquatic plethodontids that inhabit the Smokies. David Sever (St. Mary's College) has studied salamander morphology and life history, particularly of the genus *Eurycea* within and in areas adjacent to the Smokies. Some specialists, for example, Travis Ryan (Butler University) and William Gutzke (University of Memphis) have focused on unraveling the previously unknown life history and distribution of poorly studied species, in their case the Junaluska Salamander. Although many advances have been achieved using sophisticated analyses, the study of basic natural history is alive and well in the Great Smokies, if often poorly funded.

As more and more concern has been expressed because of amphibian declines and malformations in many areas of North America, studies on amphibian conservation and management are under way in the Smokies. Ted Simons (North Carolina State University), James Petranka (University of North Carolina–Asheville), and Charles (Chuck) Smith (Highpoint University) have undertaken long-term studies in order to understand how amphibian populations are influenced by current and past land use, and how amphibian populations fluctuate in abundance through time. Jim Petranka also conducted important studies tracing the effects of the toxic Anakeesta Formation on stream-dwelling salamanders, following the work of Raymond Mathews of the National Park Service in the late 1970s and early 1980s. Recent graduate students, such as Erin Hyde, Jason Lydic, and Travis Ryan, have contributed greatly to understanding salamander biology in the Smokies.

The results of the research of each of these investigators form the basis for further research on, and monitoring of, the park's amphibians. Perhaps in the future, someone will even begin a study on the frog populations of the park. Except for Petranka and Smith's study of the Wood Frog, no one has conducted research on Smokies' frogs. Information resulting from the research of these and other herpetologists who have conducted studies in Great Smoky Mountains National Park is incorporated directly into the species accounts that follow.

Ancient History and Abundant Life

The Geology and Biota of the Great Smoky Mountains

Yet it is the places that stick in my mind, which I had not seen but for Desmognathus. . . . Guyot and Clingman's Dome looming up through the mist that hung along the lines of the Smokies, when all there was between me and the Richland Balsams was miles of air and a hawk soaring.

—Emmett Reid Dunn, 1926

Great Smoky Mountains National Park extends approximately 86 km (54 miles) in straight-line distance from east to west, and varies between 16 km (10 miles) and 30 km (19 miles) in width, excluding the Foothills Parkway. Within this small area is found one of the most diverse assemblages of plants and animals in North America north of the tropics. While the focus of this book is on amphibians, visitors to the park should remember the phrase "A Wondrous Diversity of Life" as they travel its roads and trails. In order to understand why species are where they are, it is necessary to understand their life history and evolution (which was covered in the previous chapter), their physical and biotic environment (this chapter), and the influences of land use and human history upon them (the subject of the next chapter). In this chapter, I provide a brief review of the geology, physiography, and biota of Great Smoky Mountains

National Park and show how these factors affect amphibian distribution and life cycles. No one can cover these subjects comprehensively in a short field guide. Those persons interested in learning more about the physical and biotic aspects of the Smokies should consult Moore (1988), Houk (1993), or some of the numerous guides available from the Great Smoky Mountains Natural History Association.

Geology

Great Smoky Mountain National Park is located within the Blue Ridge Physiographic Province, and therefore is aligned with the older mountains to the east rather than the younger Ridge and Valley Province to the west. The Great Smoky Mountains consist of rather ancient rocks (350 million to 1 billion years old), especially in the foundation basement layer, where the rocks are composed of granite gneiss and are more than a billion years old. Such ancient rocks are exposed on the southeastern side of the park, especially along Raven Fork northeast of Smokemont and on Hyatt Ridge. The youngest rocks are located in Cades Cove, where they have been exposed through erosion. Ironically, the younger Paleozoic rocks are overlain by much older Precambrian rocks, just the reverse of what might be expected. The older rocks were thrust over the younger rocks as a result of plate tectonics. Eventually they eroded away, forming "windows" to the younger rocks underneath.

Ordovician carbonate rocks, approximately 450 million years old, are exposed in Cades Cove and Whiteoak Sink, areas where caves have developed. Most of the Smokies consist of metamorphic sandstones, siltstones, shales, and slates 600 million to 1 billion years old. These are the rocks most commonly seen within the park and form the crest of the Smokies, in part, as the Thunderhead Sandstone and the Anakeesta Formation. Table 2 provides a timeline of the important geologic events shaping the Smokies. More detailed information on the park's geology can be found in Moore (1988), Houk (1993), and in geology maps available at the park's several visitor centers.

Even in the far distant past, the geological history of the region now contained in Great Smoky Mountains National Park had a profound influence on amphibians. During the Mesozoic, amphibians evolved a rather

Table 2

Geologic History of the Great Smoky Mountains and Amphibian Evolution			
Time	Event	Result	Amphibians
1 billion–545 million years ago	sea between early continents	build up of sedimentary rocks that eventually formed the Great Smokies	
545–320 million years ago		build up of carbonate rocks	
330–250 million years ago	continents collide or push fragments together; the supercontinent of Pangaea formed	Alleghanian orogeny; land rises as vast mountain highlands	labyrinthodonts appear
245–145 million years ago	erosion; mountain building	rolling plain/hills; block mountain formation; peneplain formed; not present in Smokies	modern amphibians evolve
200 million years ago–present	Atlantic Ocean opens; erosion and weathering; regional upwarp	mountains take present configuration	plethodontids evolve
1.8 million years ago	Ice Age begins	shifting distributions of plants and animals; much of Smokies in deep freeze, although unglaciated	modern genera of amphibians appear
10,000 years ago	Holocene warming	vegetation and presumably animals assume modern distributions	amphibians populate all of the Smokies
6,600 years ago	sinkholes form in Cades Cove	large temporary pond forms at Gum Swamp	important modern amphibian breeding site

Note: Data from Moore (1988), Rueben and Boucot (1989), Houk (1993).

modern body plan. The evolution of salamanders, particularly the lungless salamanders, occurred concomitantly with the formation of the Southern Appalachians, hence the diversity that is observed today in this region.

Amphibians have had to contend with profound geologic and climatic changes throughout their phylogenetic history, yet today they are diverse, abundant, and extremely important in the functioning of ecosystems. Assuming that the salamanders, in particular, remained geographically stable within the Appalachian region throughout the Pleistocene (the Ice Age), as all paleontological evidence suggests (Holman 1995), it is reasonable to assume that they mostly moved up and down the mountains in response to climatic changes. As the vegetation communities of the Smokies assumed their present distribution in the Holocene (that is, within the last 10,000 years), it is also reasonable to assume that the majority of amphibians did likewise.

Two geological formations have had particular influence on current salamander distribution within the park. The first of these is the Ordovician limestones of Cades Cove, Whiteoak Sink, and Rich Mountain. It is only within these areas that salamanders that prefer limestone, the Cave Salamander and the Southern Zigzag Salamander, are found. No other amphibians within the park are as closely tied to particular geologic formations as are these species.

The other formation known to affect the distribution of amphibians within Great Smoky Mountains National Park is the high-elevation Anakeesta Formation. The Anakeesta comprises about 8 percent of the park and, because it has a very high sulfide content, is known as "acid rock." When mechanically disturbed and exposed to air and water, the resulting runoff, high in ferrous sulfate and sulfuric acid, can be so acidic and laden with heavy metals that salamanders, aquatic invertebrates, and fish cannot survive. For example, my wife and I could not find any living creature in the water of the spring and small, associated stream during surveys of the Smokies at Icewater Spring on Mount Kephart. Presumably, this was because the water emanated from the Anakeesta Formation. As another example, when U.S. highway 441 was modified at Newfound Gap in 1963, construction crews cut through the Anakeesta. Runoff fed into nearby Beech Flats Creek, and killed all salamander larvae and other aquatic species for 7 km (4.4 miles) down the mountain. Fifteen years later, no improvement in the stream's biota was noted in the first 2–3 km (1.25–1.9 mi) (Mathews and Morgan 1982; Kuchen et al. 1994). Toxic acidic contamination of headwater streams by the Anakeesta Formation in the Smokies and elsewhere in the Southern Appalachians also occurs naturally as a result of landslides.

Physiography

The Great Smoky Mountains National Park includes some of the highest peaks in eastern North America, with 16 reaching more than 1,829 m (6,000 ft) in elevation, as well as some rather low-elevation areas nearly in the Tennessee Valley. The highest point is 2,025 m (6,643 ft) on Clingmans Dome, and the lowest elevation is 256 m (840 ft) at the mouth of Abrams Creek. From the western side of the park to the vicinity of Clingmans Dome, the mountain crest runs east to west. From Clingmans Dome to Mount Sterling, the crest is in more of an east-northeast direction. A long spur ridge, Balsam Mountain, runs southward from the main crest of the Smokies toward Soco Gap along the Blue Ridge Parkway. Soco Gap forms a narrow connection between Balsam Mountain and the outlier Plott Balsams, whose highest peak is Waterrock Knob at 1,918 m (6,292 ft). The park is bordered on the west and southwest by the Little

Gum Swamp, Cades Cove.

Gourley Pond, Cades Cove.

Pool in stream rivulet, Cades Cove.

Tennessee River, and on the east by the Pigeon River, both of which empty into the Tennessee River drainage system. These rivers form important barriers to the dispersion of animals along the crest of the mountains. To the north lies the Tennessee Valley, and to the south and southeast are the Alarka, Plott Balsam, and, farther afield, the Great Balsam Mountains.

East of Newfound Gap, the Great Smokies are high, steep, and deeply dissected. One of the park's highest peaks, Mount Le Conte (2,010 m, 6,593 ft), is slightly isolated from the main mountain crest, and forms a distinctive physiographic feature. Other high peaks include Mount Guyot, Mount Chapman, Luftee Knob, and Big Cataloochee Mountain. Valleys are narrow and rather steeply sided. West of Newfound Gap, the ridge crest runs to Gregory Bald before descending to the Little Tennessee River. South of the main ridge line, several spur ridges (Twentymile, Welch, Forney, Noland, Thomas) run from the northeast to the southwest. In between these ridges are valleys that were heavily settled prior to the establishment of the park. West of Cades Cove in the Abrams, Cane, and Hesse Creek drainages, the hills are low and rounded, and the valleys are deeply dissected.

There are only two major valleys within Great Smoky Mountains National Park: Cades Cove and Cataloochee. Cades Cove is drained by Abrams Creek and is bordered by the crest of the Smokies to the south, Crib Gap to the east, and Cades Cove Mountain to the north. Cades Cove consists of about 3,090 ha (6,800 acres), 1,091 ha (2,400 acres) of which are open fields. It is in Cades Cove where most amphibian breeding ponds are located, particularly in Gum Swamp (sometimes referred to as Lake of the Woods), Gourley Pond, Shields Pond, and in the shallow wetlands bordering Abrams Creek. Additional breeding ponds and shallow wetlands are located in several sinkholes along Finley Cane Trail in Big Spring Cove (just east of Crib Gap), in the Sugarlands area, and along Cane Creek at the Park's western edge. Sinkhole ponds are particularly important as amphibian breeding sites, and some may have persisted for at least 6,600 years (Davidson 1983).

Cataloochee Valley is much narrower than Cades Cove, at most being only about 300 m (984 ft) at its widest. No farm ponds remain, if any ever were present. In Cataloochee Valley and adjacent Little Cataloochee Valley, wetlands used by amphibians for breeding are found in shallow depressions and seeps on the valley floor, as well as in an abandoned trout hatchery pond, now drained. Elsewhere in the park, human-made breeding "ponds" are found along the road to Tremont, at an artificially constructed farm pond south of the Methodist Church and the sewage treatment pond in Cades Cove, in the spring house at Cataloochee near the present ranger station, and in canals in Big Cove. Beaver ponds form important amphibian breeding sites in Big Cove and Bone Valley on the park's North Carolina side. Undoubtedly, additional small woodland ponds and wetlands used by amphibians will be found, for example, along lower Hazel Creek, Panther Creek, Indian Creek, and in Smokemont.

The Great Smoky Mountains can be divided into 25 major watersheds (King 1939b), nearly all of which empty eventually into the Tennessee River system. Nearly 960 km (600 miles) of streams are found within the park. Most streams are clear and cold, even in summer, and in times of high water can be transformed into raging torrents. In late summer, however, the same streams may have such reduced flows that they are easy to cross on exposed rocks without touching water. Nearly all stretches of all streams in the Smokies provide habitat for salamanders, the exception being high-elevation headwater trickles contaminated by the Anakeesta.

Medium-sized stream, Fighting Creek.

Climate

A number of different climates are found within the Great Smoky Mountains, ranging from the warm, dry, temperate regions around Chilhowee Reservoir in the west to the boreal Canadian Zone climates of the highest peaks, such as Clingmans Dome and Mount Guyot. In general, as one goes up approximately 304 m (1,000 ft) in elevation, the climate is similar to an area 400 km (250 miles) to the north. Thus, the climate on top of Clingmans Dome is similar to the climate approximately 1,800 km (1,125 miles) to the north of the Smokies, that is, at the southern end of St. James Bay, Québec. Such a range in climate allows many types of animals to occur near to one another in an area where they never would naturally come into contact based on life history or physiological requirements alone. In the Great Smokies, northern boreal forest animals can be seen on the same day as animals whose origins are in the subtropics simply by climbing the mountains.

Temperature

As elevation increases, temperature decreases in the Great Smokies. As a rule, the temperature decreases 1.67°C (3°F) for every 304 m (1,000 ft). Thus, even on the warmest summer days, the tops of the mountains are relatively pleasant, averaging about 18°C (65°F) for a high on Clingmans Dome in July. Winter is another matter, however, as the mean high temperature is only 1.7°C (35°F) in January on Clingmans Dome. In contrast, the mean January high temperature in Gatlinburg is 10.6°C (51°F), whereas the mean high in July is 31°C (88°F). The potential for a wide range of temperatures based on topography influences amphibian distribution and behavior. Many salamanders and frogs are active only during cool weather. Such species, therefore, are usually found only in the lowlands and only during the spring and autumn, when temperatures are cool. One example is the Upland Chorus Frog, which is active only during a late winter to spring breeding season, and then virtually disappears. Southern Red-backed Salamanders are active in the spring and autumn at lower elevations, moisture permitting, but disappear in the lowlands during the warm summer months, even during periods of above average rainfall. However, at cooler high elevations, these salamanders may be active throughout the summer.

Certain high-elevation salamanders never experience warm temperatures and, indeed, cannot tolerate them when artificially exposed to warm thermal environments. Jordan's Salamander can nearly always be found on the highest peaks in summer, but does not venture below 775 m (2,545 ft). Its ecological and, perhaps, physiological requirements do not allow it to descend to lower elevations, even during the cool weather of spring and autumn. Clearly, temperature influences distribution and behavior, at least in this and perhaps other species, but it is not the only consideration that affects these attributes.

Likewise, terrestrial salamanders from the warm lowlands rarely venture into the highlands, although as in the case of the Southern Red-backed Salamander, flexibility sometimes occurs in the timing of activity at different elevations, presumably in conjunction with temperature effects. Because some amphibians, such as American Toads, appear to tolerate warm temperatures reasonably well as adults, perhaps temperature is more important for them in controlling egg and larval development. In any case, certain amphibians prefer warm temperatures, whereas others prefer cool temperatures. The range in elevation within the Smokies allows both groups to occur within a relatively small geographic area, despite their differences in preferred environmental temperatures.

Precipitation

In addition to overall climate and temperature, the range of elevation and the way in which high ridges are oriented within the Smokies affects precipitation patterns. The greatest amount of precipitation falls at the highest elevations, as clouds swirl across the knife-edged ridges. Rain and snow are frequent on the high peaks, as are the mists and fogs, which leave the vegetation wet or damp, even when rain is not falling in the lowlands. The higher elevations also have higher humidity than the lowlands, allowing salamanders, in particular, ample opportunity to forage in the leaf litter and on wet bark in thickly forested coves. For comparison, Gatlinburg receives an average of 137 cm (54.1 in) of precipitation per year, and 23 cm (9 in) of snow. Precipitation falls on an average of 99 of days per year. In contrast, Clingmans Dome receives 208.5 cm (82.1 in) of precipitation per year, and 213 cm (84 in) of snow. There, precipitation falls an average of 126 days per year, and this does not include the water that adheres to vegetation as condensation during cloud cover, mists, and fogs.

Dry oak-hickory-pine forest, Tunnel Ridge.

Even taking into account the effects of elevation, precipitation does not fall evenly within the Great Smokies. Storms blowing in from the northwest tend to dump their rain and snow on the northern side of the park and on the highest peaks. As the winds cross the mountains, they drop less precipitation on the southern (North Carolina) side of the park, creating something of a rain shadow along the mid and lower slopes, such as on the south-facing ridges from Eagle to Deep Creeks. On the other hand, tropical depressions and the remnants of hurricanes may approach from the south, albeit infrequently. When this happens, torrential downpours loosen high-elevation vegetation rooted shallowly over bare rock, and landslides cascade down the mountain flanks. Several V-shaped scars from such events are visible near Newfound Gap.

In addition to the southern rain shadows, the relative lowlands in the extreme western side of the park also are very dry. The ridges overlooking Cane and Abrams Creeks turn to tinder during the summer months, and it is here where natural periodic lightning-generated fires once burned, giving rise to an oak-pine fire-maintained vegetation community. In this western region and throughout much of the lowlands, most terrestrial salamander activity occurs only during the cool, wet spring and virtually ceases during the summer. Salamander species richness also is much lower in the dry oak-pine woodlands, although frogs are more common around the few wetlands and creeks than they are at higher elevations.

The timing of precipitation also affects amphibian activity in the Smokies. Fortunately, much of the rain in the lowlands, in particular, falls from late winter through early summer, when ponds fill and pond-breeding amphibians commence reproductive activity. As the year advances, summers become increasingly dry in the lowlands. At the higher elevations, thunderstorms rising from the hot valleys frequently buffet the mountain ridges and dump heavy rainfall on the highest peaks. Thus, higher-elevation species enjoy relatively stable precipitation throughout the spring and into the autumn. The stable warm season precipitation patterns allow high-elevation salamanders to be active from early spring to late autumn, depending on when the temperatures start dropping below freezing. When rains, associated with cool weather fronts, arrive in the late autumn to early winter, lowland salamanders again resume some activity before winter sets in. In mild winters, some Smokies amphibians remain active throughout the season at lower elevations.

All of these events are relative, since they depend on weather in the Smokies that is, as elsewhere, often unpredictable. In some years, a normal spring rainfall is followed by an abnormally severe summer drought. When this happens, breeding ponds sometimes dry before larvae have a chance to metamorphose. In other years, the spring rains simply fail to materialize, and pond-breeding amphibians must forgo a year of reproduction. Even in "normal" years, some low-elevation streams and rivulets dry up before they can enter larger creeks; in severe droughts, long sections of dry streambeds appear, seeps and wet rock faces disappear, and vegetation withers. Presumably, amphibians wait out the dry periods deep underground where moisture levels are favorable. In wet years, however, amphibian abundance soars. Pond-breeding species then have ample breeding sites and a sufficient hydroperiod in which to complete metamorphosis. Terrestrial species have free reign to forage and mate, and their invertebrate prey base is abundant. Rainy years probably are extremely important in amphibian population dynamics. Reproduction that occurs during wet years helps to offset poor survivorship during years when precipitation is deficient or occurs outside the reproductive season.

Vegetation Communities

There are five major forest communities within the Great Smoky Mountains National Park, although 80 percent of the park falls within the Eastern Deciduous Forest Ecosystem (Houk 1993). Some botanists have further subdivided the vegetation into as many as 67 florally distinct communities. The forest communities of the mountains acquired their present vegetative composition and distribution about 10,000 years ago (part of the Holocene) after the retreat of the last Ice Age. Before Holocene warming, the mid to high peaks of the Smokies were devoid of trees and shrubs and were inhabited only by the hardiest of tundra plants. The boreal forest was located on the lowest slopes and, even with warming, tree invasion of higher slopes was inhibited by boulders falling from the rocky peaks still subject to intense fracturing during freeze-thaw cycles. Today, there may be more than 130 species of trees in the five major forest communities (Kemp 1993) and more than 4,000 species of vascular plants.

No one species of amphibian is associated entirely with a single forest community, although some of the high-elevation salamanders (for example, Jordan's Salamander, Ocoee Salamander, Pigmy Salamander) are more often found in the spruce-fir community than in other community types. As previously noted, this distribution may have more to do with the environmental requirements of the species rather than with the species of trees found in the dominant community. Habitat structure, particularly one that retains moisture and high humidity, is important in shaping salamander distribution. The high-elevation coniferous forest appears ideal in providing shade, cover (in the form of coarse, woody debris), and abundant surfaces for moisture condensation. The present-day high-elevation salamander assemblage probably colonized the rocky peaks of the Smokies, uninhabitable throughout the Pleistocene, during the Holocene as the forest crept up slope with the changing environmental conditions.

The spruce-fir forest is dominated by Red Spruce *(Picea rubens)* and Fraser Fir *(Abies fraseri),* and is found generally above 1,676 m (5,500 ft), although the community descends to 1,372 m (4,500 ft) in some locations and individual Red Spruce are found at still lower elevations. Unfortunately, nearly 95 percent of the Fraser Firs have been killed by the balsam woolly adelgid, and their dead trunks are seen throughout the highlands. This is the Canadian Zone boreal forest of high moisture, cool or

Coarse woody debris in cove forest, Roaring Fork.

cold temperatures, and high humidity. The ground surface is often dense with fallen tree branches and trunks and carpeted by thick layers of tree needles. The wet, rotten, woody debris and dense needle mats provide ideal hiding places for terrestrial salamanders. Streams arise in this habitat, and usually begin as small seeps and springs. As they trickle through the dark green forest, they gather momentum. Even at higher elevations, aquatic salamanders, particularly duskies *(Desmognathus)* and Blue Ridge Two-lined Salamanders, may be plentiful within the headwater streams.

At somewhat lower elevations (1,067–1,524 m [3,500–5,000 ft]), deciduous northern hardwoods predominate. This community is dominated by Sugar Maples *(Acer saccharum)*, American Beech *(Fagus grandifolia)*, and Yellow Birch *(Betula alleghaniensis)*. Many terrestrial and aquatic salamanders reach their lower or upper distributional range within this community; frogs are scarce and always transient. Cove hardwoods, the third community, constitute the most diverse forest community in the Smokies, one that is endemic to the Southern Appalachian Mountains. It occurs generally below 1,372 m (4,500 ft) in sheltered valleys. This community is dominated by Tulip Poplar *(Liriodendron tulipifera)*, Dogwood

(Cornus florida), Red Maple *(Acer rubrum)*, Sweetgum *(Liquidamber styra-ciflua)*, White Basswood *(Tilia americana* var. *heterophylla)*, Yellow Buck-eye *(Aesculus flava)*, and Black Birch *(Betula lenta)*. Both hardwood com-munities have complex understory vegetation, often with much coarse, woody debris, which provides cover for terrestrial salamanders. The streams through these hardwood forests are rocky and fast paced, and salamanders are very common along streamsides and in the water.

Two somewhat specialized forest communities are found in the Smo-kies. The hemlock community is dominated by Eastern Hemlocks *(Tsuga canadensis)*, commonly called "spruce-pines" by natives of the southern mountains, and is common between 1,067 m (3,500 ft) and 1,524 m (5,000 ft) in elevation. Hemlocks descend to much lower elevations along the cold mountain stream valleys and overlap considerably with both hard-wood forests and the spruce-fir forest of the higher elevations. Hemlocks are massive with tall, straight trunks. When they fall, they provide excel-lent habitat for salamanders, both in the rotting wood and under exfoliat-ing bark. Eastern Hemlocks are now threatened by the hemlock woolly adelgid, another introduced exotic species.

The pine-oak forest occupies the dryer areas of the park, particularly the area west of Cades Cove and at mid-elevations on the North Carolina side of the park. This forest is dominated by Southern Red *(Quercus falcata)*, Northern Red *(Q. rubra)*, Scarlet *(Q. coccinea)*, Black *(Q. velutina)*, and Chestnut *(Q. prinus)* Oaks, and by Pitch *(Pinus rigida)*, White *(P. strobus)*, and Table Mountain *(P. pungens)* pines. Soils are dry, as is the leaf litter. Prior to human intervention, this community burned frequently in the western regions of the park, and a fire-adapted vegetation community resulted. Terrestrial salamanders are few, and they are usually found only during the cool, wet times of the year. Aquatic salamanders and frogs are found along streamsides, where they likely remain close to water. The bottomlands along Cane Creek and Abrams Creek likely formed a corri-dor from the Tennessee Valley into Cades Cove. As a result, amphibian species richness is surprisingly high, particularly for frogs.

The Lost Chestnut Forest

The American Chestnut *(Castanea dentata)* was once the dominant tree in the hardwood forests of eastern North America. The tree was extremely important to both wildlife and humans, providing abundant food, shelter, and building material (Houk 1993). The dominance of the tree in forest ecology gave rise to the name Chestnut-Oak forest as an ecological association. In the Great Smokies, the American Chestnut covered 31 percent of the land, roughly 72,348 ha (159,165 acres). Beginning about 1925 and 1926 in the Great Smoky Mountains, an exotic fungal blight *(Cryphonectria parasitica),* introduced from Asia in 1904, devastated the forest and killed virtually every tree. By 1929, 99 percent of the trees were infected, although it took 10 years for a large tree to die. Today, chestnut stumps still sprout and give rise to saplings that succumb to the blight before reaching reproductive age. Massive stumps and fallen trunks decay slowly, and they can still be found in Great Smoky Mountains National Park. Today in the Smokies, the Chestnut-Oak forest has been replaced by a forest dominated by red maples and northern red and chestnut oaks (Woods and Shanks 1959). Unfortunately, the replacement forest did not help the several species of moths that became extinct because their host chestnuts were no longer present.

One day, my wife and I collected salamanders through a grove of chestnut stumps, some of which reached 2–2.5 m (6–8 ft) high, adjacent to Parsons Branch Road along Rabbit Creek. It must have been an incredible forest. Aside from depriving terrestrial salamanders of abundant cover, it is tempting to speculate what the loss of this dominant tree might have had on the amphibian community. Jim Petranka recalls Roger Barbour's stories of how Green Salamanders could be found commonly under chestnut bark. Barbour, a professor of zoology at the University of Kentucky, spent countless hours roaming the hills of eastern Kentucky, and he wondered out loud if Green Salamanders were really a bark crevice species, rather than a rock crevice species as currently envisioned. Did the loss of the American Chestnut tree cause declines in this salamander's population? No one, of course, can ever say for sure. The only Green Salamander ever found within the Great Smokies was discovered under a log of unreported species.

Biota

There are a great many types of microbes, fungi, plants, and animals in Great Smoky Mountains National Park: soil microbes and tiny worms, mushrooms, shrubs, annuals and perennials, insects, spiders, millipedes,

Milas Messer place, Cove Creek, North Carolina, 1937. Messer operated a tanning business in addition to the farm. Photographer E. E. Exline. GSMNP III-F-12632.

Grazing area west of Spence Field, 1934. The open woods and lack of understory was the result of grazing. Photographer E. E. Exline. GSMNP III-F-12609A.

shore. The large human population impacted the land to a considerable extent, judging from the hundreds of photographs in the archives of the National Park Service at Sugarlands.

Reforestation occurred rapidly after the park was established and the National Park Service acquired ownership, however. Sufficient source populations of amphibians must have remained in the fragmented landscape to recolonize formerly occupied terrestrial and aquatic sites. After nearly 70 years, the amphibian community may be approaching what it was like before human settlement, although it is clear that settlement has had long-term effects on the park's terrestrial salamander communities (Hyde and Simons 2001). Subtle differences undoubtedly remain in terms of distribution, species richness, and abundance. Today, it is sometimes difficult for visitors to discern where people had cleared and farmed; only certain horticultural plants or the arrangement of stones into piles or fences hint at a past human presence. Where once springs were dug or enlarged, salamanders now hide under stones and planks, as they should. Large-scale commercial timbering, however, had a much greater impact on amphibians than small-scale farms.

Commercial Timbering

In the 1800s, settlers cleared the land and cut trees for local consumption. By 1884, the Smokies already had 100 small sawmills. These operations relied on selectively cut timber, particularly the valuable Cherry, Black Walnut, Ash, and American Chestnut. Individuals and companies soon discovered that the chestnut bark contained tannins, which opened up additional commercial possibilities, since the tannins were used in the process of tanning leather hides. In the mountains surrounding Cataloochee, foresters estimated that 80 percent of the harvestable timber was in chestnut and oaks in the early 1900s.

As northern timber companies exhausted forests in the Northeast and in the Great Lakes region, they began to turn their attention to the Southern Appalachians. In this, they were spurred on, in turn, by 1901 U.S. Department of Agriculture estimates of the immense forest ready for cutting in the Great Smokies; by the cheap, non-union labor and land prices (anywhere from 50 cents to 10 dollars per acre, depending on location); and by technological improvements. Eight large timber companies began buying lands from small landowners and speculators. During the

Lumber mill on Big Creek, date unknown. This huge mill operated where the picnic area and campground on Big Creek are today. Note the large mill pond. GSMNP III-L-17810.

course of the timber boom over the next three decades, two-thirds of the Great Smokies would be clear-cut, huge lumber mills would be constructed, 333 km (200 mi) of logging railroads would be built, and thousands of men would be employed. The land has not yet recovered, and perhaps never will in a time frame meaningful to humans.

The corporate timber companies constructed roads, railroads, processing mills (complete with large ponds), and both relatively permanent and temporary settlements. Logs were cut, put on railroad cars, and hauled to the mills, either locally, such as at Big Creek, or outside the park, such as at Townsend, for further cutting and finishing. The primary way logs were cut was through clear-cutting large sections of forest. Logs were skidded, carried by tramlines, or put in logging chutes for quick dispatch down the mountainside. Overhead skidders could reign in logs from a mile away. In some places, splash dams were built across streams, and logs were floated in the pond until it was filled with them (see photo in Schmidt and Hooks 1994:6). Then the splash dam was burst, and the logs cascaded down the streambed in an avalanche, scouring everything in their path. On Hazel

Creek, letting splash dams loose killed so many trout that residents walked along the sides picking up bucketfuls (Brown 2000). Heavy log removal created a "veritable arroyo of torn shores and skimmed stones out of mountain streams" (Brown 2000:25). The mountains were literally "skinned" as logs moving on skidders, in particular, destroying everything in their path. The level of destruction can only be appreciated by looking at archival photographs (see Weals 1993; Schmidt and Hooks 1994; Brown 2000).

After the logs were cut, there was no attempt at reforestation. Instead, a tremendous amount of slash and woody debris remained, which quickly dried out when exposed to the direct sun. This left the land vulnerable to fires, and vast stretches burned as a result of lightning strikes and from sparks thrown out from the locomotive engines. Sometimes fires burned for days, and the Great Smokies experienced their worst fires in the 1920s and 1930s. A large fire in 1924 burned from Yellow Creek and Sinking Creek to the top of Mount Guyot, a straight-line distance of 4.3 km (2.6 mi). After the fires, of course, came floods. With the surface vegetation and root structure effectively removed, water ran in torrents down the sides of the mountains, carrying topsoil and humus downward and silting low-elevation streams and rivers. Native ground vegetation and wildlife were simply left to fend for themselves, and many herbaceous plants, ferns, flowers, lichens, and other vegetation never recolonized the clear-cuts. Formerly cold streams were exposed to direct sunlight, and native trout disappeared, eventually leading to the stocking of nonindigenous western trout.

The boundary between areas that were clear-cut more than 70 years ago and areas that were not cut can still be readily seen in some places within the Great Smokies. After walking the 2.9 km (1.73 mi) from the Clingmans Dome road down the Noland Divide Trail, a hiker will come to an area where the clear-cut line is quite visible. On one side of an imaginary northwest-southeast line is a spruce-fir forest; on the other is a deciduous woodland populated mostly by even-aged, spindly oak and maple trees. The habitat change is dramatic, despite the fact that the area was cut long ago. Likewise, the terrestrial salamander fauna changes. Salamanders

Facing page top: Lumbering operations, Mount Cataloochee, date unknown. Note the railroad tracks and the huge amount of slash. GSMNP III-L-7191. Facing page bottom: Railroad cut above Flat Creek, North Carolina, with Mount Sterling in the background, 1931. GSMNP III-L-16466.

Slash left after clear-cutting by the Little River Company near Elkmont, date unknown. Note the skidding cable. After the slash came the fires and erosion. GSMNP III-L-7192.

are less abundant in the oak-maple forest, and some species common in the spruce forest, such as the Pigmy Salamander, do not venture into the oak-maple forest. The Southern Red-backed Salamander, normally considered a species of the warm, low to middle elevations, also reaches its highest point within the park in the oak-maple forest along this trail.

Biologists have debated the effects of clear-cuts on Appalachian salamanders for about a decade now. Because studies differ in scope, sampling techniques, and areal extent, it is possible to draw somewhat different conclusions among studies. No one doubts, however, that clear-cutting in the southern mountains has drastic and immediate consequences for terrestrial salamanders: their numbers decrease dramatically. Salamanders in mature forest in western North Carolina are as much as five times more abundant than they are on clear-cuts, with twice as many species occurring in mature forest as on clear-cuts. It may take 40 to 80 years, perhaps even longer in dryer conditions, for salamander populations to recover from clear-cutting in the Southern Appalachians (Petranka, Eldridge, and Haley 1993; Petranka et al. 1994). The large variance results from the type of cutting, the extent of soil disturbance, whether fire and erosion had impacted the site, the distance from unaffected areas that might be a source of new colonizers, and a host of other considerations.

Adverse long-term impacts of logging recently have been corroborated by intensive research on the salamanders of the Mount Le Conte USGS topographic quadrangle in Great Smoky Mountains National Park. Here, Hyde and Simons (2001) assessed terrestrial and streamside salamander abundance and species diversity using a variety of sampling techniques over several years. They carefully selected areas that had been disturbed by logging or agriculture, and compared them with areas that had not been disturbed. Their results showed that disturbance significantly negatively impacted populations, and that salamander diversity and abundance were still being affected by land-use practices that ended 60 years ago. These results mirror the findings of biologists working on other taxa in the Great Smokies, where only half the species and one-third of the forest canopy have recovered to precut conditions in areas that were intensively logged (Brown 2000).

Dams and Reservoirs

A major change affecting habitats adjacent to the Smokies resulted from the construction of dams along the Little Tennessee River west and southwest of the park. Alcoa Aluminum built Cheoah Dam in 1919 and Calderwood Dam in 1930. These dams effectively blocked the upstream dispersal of certain aquatic salamanders in the Little Tennessee River, such as Common Mudpuppies and Hellbenders, and eliminated habitat for other species within the valley of the Little Tennessee. The reservoir behind Chilhowee Dam, built between 1955 and 1957, further inundated 794 ha (1,747 acres) downstream of the previous dams, including the mouth of Abrams Creek. The subsequent dam and reservoir prevented recolonization of Abrams Creek by rare native fishes and Hellbenders, eliminated from this drainage in 1957 in a scheme to improve trout fishing. Chilhowee Reservoir also inundated the former town of Chilhowee, where several species of amphibians (Northern Cricket Frog, Northern Leopard Frog), now apparently absent from the park, once were found.

The huge Fontana Dam, 146 m (480 ft) high, created Fontana Lake, a 5,318 ha (11,700 acre) reservoir with 400 km (240 mi) of shoreline. Fontana Dam is the tallest dam east of the Rocky Mountains. During its construction between 1942 and 1945, more than 6,300 workers were employed in often makeshift communities near the construction areas. Many local communities were displaced, and the agriculturally productive valley of the Little Tennessee River disappeared. With it, so did the large river fish, such as paddlefish and lake sturgeon, and of course the large river salamanders. Native fishes and amphibians disappeared from the lower reaches of Twentymile, Hazel, Eagle, and Forney Creeks, and populations of Hellbenders became confined to Deep Creek, at least within the park. The reservoir probably eliminated the rare Junaluska Salamander from this portion of its range, although it still occurs south of the park along creeks in the Cheoah River valley.

The reservoirs created by this series of dams isolated surviving populations of lowland, aquatic, stream-dwelling amphibians and prevented streamside dispersal. Streams, shallow wetlands, and forests were replaced by a rather sterile, deep water environment. Fontana Lake also contains high concentrations of manganese, copper, zinc, and mercury. Because the lake levels fluctuate dramatically, reservoir shorelines, such as those on the north shore of Fontana Lake, have never stabilized in terms of vegetative cover. During times of low water, the shorelines consist of widely exposed borders of rock and mud. The deep drop-offs do not provide amphibians with proper shallow-water breeding sites, and the lack of shoreline vegetation prevents a safe approach in a thermally favorable environment. Still, some amphibians, such as the American Toad and American Bullfrog, find breeding sites in the inlets formed from drowned streambeds. Today, no amphibians live within the reservoirs themselves.

Park Development

Great Smoky Mountains National Park became a unit of the National Park Service in 1934, although the park was not officially dedicated until 2 September 1940. At this time, the Smokies were anything but an untamed wilderness. Timber companies had worked night and day trying to cut as much wood as possible before the National Park Service took over, with no regard for reclamation or the detrimental impacts of last-second logging. There were hundreds of small farms scattered throughout the valleys and coves, and towns, such as Proctor on Hazel Creek, had yet to be dismantled. Indeed, the park would not take on its present configuration until the mid-1940s, when the private land isolated by the creation of Fontana Lake was added—at times over vigorous opposition (Brown 2000). Scattered old-growth forest remained, however, particularly at higher elevations and in some of the difficult-to-access coves. There were also inholdings, such as the cabins at Elkmont and the 46.4 ha (102 acre) estate of Louis Voorheis on Twin Creeks, retained until 1952. In 1933 while touring the area which became the park, NPS employee Joe Manley called it "the most fragmented ecosystem" he had ever seen (Brown 2000:113).

Early park supporters clearly had different visions of what Great Smoky Mountains National Park should be (see Brown 2000 for an extensive review). For example, David Chapman of the Appalachian Club wanted to build airplane landing fields in Cades Cove, Cosby, and Greenbrier, and a huge lake for fishing in Cades Cove. He even tried to convince the Tennessee Valley Authority to build an 18 m (60 ft) dam on Abrams Creek to allow boating. Others wanted golf courses, in order to set a precedent for constructing them in national parks, and even ski lifts. Fortunately, none of these schemes was approved, thanks to the vigilance of both local and national organizations. Instead, commercial development has rapidly altered the landscape of communities surrounding the Great Smokies, and nearly isolated the park.

Photographs from the 1930s and1940s dramatically show the devastation of certain sections of the Great Smokies (Schmidt and Hooks 1994; Brown 2000). As NPS took control of the land, towns and buildings were dismantled, roads allowed to fall into disuse, and the land gradually came to reclaim itself. Cades Cove changed from a valley of small farms into a valley with subsidized ranching. With the elimination of much human presence, a portion of the vegetation and wildlife returned. Whereas white-tail deer and bear had become scarce, they re-established themselves in abundance. Apparently, the amphibians did likewise, judging from the presence of many species today. There was no management directed specifically at the recovery of amphibians or their habitats, as NPS worked on visitor facilities and interpretation. In this regard, NPS received a tremendous boost with the coming of the Civilian Conservation Corps (CCC). CCC operated in the Smokies from 1933 to 1942. There were 17 CCC camps, employing 4,350 men, which often were situated on the old logging camp sites. The men built trails, bridges, and buildings, improved roads, dismantled settler structures, planted trees, and built and operated trout hatcheries. They also constructed a 1,000 m (3,280 ft) ditch in Big Cove as part of a nursery connected with a revegetation program, resulting in significant alteration of the cove's low wetlands.

Although the development of park access and tourist facilities undoubtedly affected amphibians, the impacts were minor compared with the previous 100 years of land alteration. The same cannot be said for other schemes. In 1946, the Soil Conservation Service (SCS, now the Natural Resources Conservation Service) initiated a "historical" program to perpetuate the open scenery in Cades Cove, at least as it was when the park was dedicated. As part of this program, stream channels were straightened and trees and shrubs along the banks were cleared in order "to provide quick drains and to prevent 'unnecessary flooding.'" (Brown 2000:203). In 1964, SCS undertook further "channel improvement" in Cades Cove. They again removed riparian vegetation and eliminated all the plants, logs, and limbs that kept erosion in check. As a result, agricultural effluent from the ongoing cattle ranching operations quickly entered Abrams Creek and traveled rapidly downstream. In 1980, high levels of fecal coliform could be detected as far as 15 km from Cades Cove.

Not all wetlands projects begun since the park's development have had negative consequences for amphibians. For example, a farm pond constructed in 1966 south of the Methodist Church on the Cades Cove loop (hereafter referred to as Methodist Church Pond), presumably to water cattle, provides excellent breeding habitat for Spotted Salamanders, Eastern Red-spotted Newts, and Northern Green Frogs in an area where ponds are scarce. Amphibians also breed in the backwaters of small check dams in Cades Cove, and in the highly nutritive Cades Cove sewage treatment pond, a favorite of Cope's Gray Treefrog.

The alteration of wetlands within Cades Cove and elsewhere undoubtedly had major impacts on the diverse amphibian community within their respective areas. Small, temporary ponds are crucial as amphibian breeding sites, and draining and ditching such wetlands eliminated their availability for use by amphibians. Perhaps the elimination of such wetlands accounts for the scarcity of some species of amphibians within the cove today, such as the Mole Salamander, Northern Leopard Frog, and Eastern Spadefoot. The modification of wetlands in Cataloochee and Big Cove also could have had detrimental impacts to amphibians, but it is difficult to say what the impacts were without knowing much about the amphibian community that existed at these locations prior to alteration. The National Park Service is currently undertaking a wetlands restoration program in Cades Cove, Oconaluftee, and Cataloochee that should benefit

Farm near what is today Loop A of the Elkmont campground, 1918.
GSMNP III-L-4751.

many species of amphibians. Each area previously modified has been revisited and mapped, with recommendations developed for facilitating a return to more natural wetland conditions.

With park development also came changes to the terrestrial environment. As vegetation succession occurred, areas that were once open fields, pastures, or gardens, became forested. When looking through the park's photographic archives, it is often difficult to correlate the photo in hand with the way the area appears today. Changes in the environment through vegetation succession undoubtedly account for some of the changes observed in species' distribution in the Great Smokies. For example, there are specimens in the park's research collection of the Upland Chorus Frog and Spotted Salamander from Elkmont, where they do not occur today. Prior to the mid-1930s, Elkmont consisted of small farms, a hotel and cottages, and a timber camp at various times. The area was open canopied, with many fields and, presumably, small wetlands. Today this area is completely reforested. The shallow grassy wetlands and small depression ponds favored by Upland Chorus Frogs and Spotted Salamanders no longer occur there. Likewise, reforestation of the Cane Creek drainage may account for the absence of Northern Leopard Frogs and other species from this region.

Roads and Tourist Facilities

There are currently 397 km (238 miles) of paved roads and 243 km (146 miles) of unpaved roads, within Great Smoky Mountains National Park. The heaviest traveled roads are U.S. Highway 441 from Gatlinburg to Cherokee, and the Cades Cove Loop Road in Cades Cove. The roads generally follow old settler roads, which in turn followed Indian or game trails. A number of roads, such as roads on Balsam Mountain and along Little River, follow the old railroad logging tracks. There are exceptions. The Roaring Fork motor loop was completed in 1963 to allow visitors to Cherokee Orchard a quick return to Gatlinburg. Roads are important to visitors in the Great Smokies, so much so that in 1968, a National Geographic Society article stated that "only 6% of the visitors actually took to the trails on foot" in calling the Smokies a "drive-in" park (Brown 2000:175). Stuck in a traffic jam along the road to Newfound Gap, one gets the feeling that the situation has not changed much.

The most serious incident involving roads and their effects on amphibians occurred when U.S. Highway 441 was redesigned at Newfound Gap in 1963, and construction workers cut through the Anakeesta Formation. Runoff fed into nearby Beech Flats Creek and killed all salamander larvae and other aquatic species far down the mountain. Fifteen years later, no improvements in the stream's biota were noted in the first 2–3 km (1.25–1.9 mi) (Mathews and Morgan 1982; Kuchen et al. 1994). Concern also has been expressed about the potential effects on aquatic species should the road along the north shore of Fontana Reservoir ever be built. Anakeesta rock is exposed in this area and subject to frequent landslides. Any construction cutting through Anakeesta Formation rocks would be cause for concern since all downstream aquatic fauna would be immediately jeopardized.

Existing roads are well maintained by the National Park Service. Although much habitat was disturbed when the roads originally were constructed, every effort has been made to minimize further impacts, such as from silt runoff into streams. Vegetation is kept clear of rights-of-way, but de-icers, such as road salt, are not used to hasten snow melt. Salt is thought to have detrimental impacts on amphibians breeding near roads, although surprisingly little data are available as verification. The main threat to the park's amphibians from current roads is highway-related mortality. Frogs and salamanders are often killed in substantial numbers along Little River Road and on U.S. 441 on warm, wet, spring and summer nights. Drivers may be unable to even see slender salamanders, and frogs moving quickly across roads may be inadvertently run over. There are other drivers, however, who seem to delight in crushing a toad or treefrog as it sits in the road. Such people are beyond contempt.

Fortunately, road traffic decreases greatly at night when most amphibians are active on the ground surface and when major breeding migrations take place. The road most likely to have some potential adverse impact on amphibian populations is the Cades Cove Loop Road, which is closed from dusk to dawn. Nightly closure is critical, because some of the park's largest and most important amphibian breeding sites are adjacent to its 18 km (11 mi) circuit. Unless breeding migrations are stimulated under gray, gloomy, rainy spring afternoons, however, most of Cades Cove's unique amphibians are relatively safe from highway mortality. Ecopassages (sometimes termed "toad tunnels" when built for amphibians) have been used successfully in several parts of North America and Europe to facilitate amphibian migration and movement across dangerous roads. As long as Cades Cove Loop Road, with its overwhelming car traffic, remains closed at night, there are no places within Great Smoky Mountains National Park where ecopassages currently are necessary to benefit amphibians. On a more positive side regarding roads, biologists occasionally have resorted to "road cruising" as a method for sampling the park's amphibians, and even traffic-killed specimens have been preserved as museum vouchers or for toxicological research.

In 2001, more than 9,457,000 persons visited Great Smoky Mountains National Park. The Great Smokies were not always so popular, at least prior to the mid-1960s. Administrators within the National Park Service began to advocate that a large number of tourists be used as a measure of successful park management, an approach that emphasized facilities at the expense of wildlife and ecosystem management and research (Brown 2000). Tourism was promoted with good results. Between 1964 and 1975, the park experienced a 63 percent increase in visitation, 33 percent increase in hiking, and a 63 percent increase in horseback riding. As previously, most visitors today enjoy brief stays, concentrating their activities along U.S. Highway 441, on the Little River Road into Cades Cove, and on the Roaring Fork motor trail. A total of 957 camping sites are available, including substantial concentrations in Smokemont, Elkmont, and in Cades Cove. More than 1,333 km (800 mi) of trails, both for foot and horse travel, allow visitors to reach nearly all sections of the park. Trail erosion (see Bratton, Hickler, and Graves 1979), litter, waste disposal, automobile-related pollution, and the sheer wear and tear of so much visitation take their toll. Even as early as 1942, the Smoky Mountains Hiking Club complained about litter. Although the number of visitors is still touted by the National Park Service, it has paid more attention to natural resource planning and management in recent years.

Undoubtedly, the cumulative impacts of all these visitors affect amphibian habitats. There are no studies to assess how visitation, even from obvious sources such as highway mortality, impacts amphibians within the Great Smokies. Fortunately, most current visitor-related impacts probably are localized and somewhat contained, at least from a small animal's perspective. Research needs to address ways to minimize visitor influences on amphibians and other biota, especially in the siting of visitor facilities and on the effects of roads.

External Threats

Pollution, Acid Rain, and Ozone

One of the most serious threats to the ecosystems of the Great Smokies originates from sources hundreds of miles away, as well as from the thousands of vehicles within or adjacent to the park: atmospheric pollution.

Wind patterns push the pollution from midwestern North America south-east to the Smokies, where it concentrates and precipitates with increasing elevation. Even forest fires burning in Mexico in May 1998 made driving in the Smokies difficult because of the dense smoke blanketing the mountains. Thus, the air is paradoxically often cleaner in the metropolitan lowlands than it is in the high elevations of the National Park. Much of this pollution is invisible to the eye, although even this is changing. In past years, clear rolling, blue ridges could be seen far into the distance, but now it is sometimes impossible even to see the next ridge only a few miles away. During the 1990s, the average visibility in the Great Smokies decreased 80 percent in the summer and 40 percent in the winter. With the pollution comes acid rain. One measurement of rainfall acidity at Twin Creeks recorded a pH of 3.2 in 1978 (neutral is 7.0; the lower the pH, the more acidic the liquid becomes). Currently, the rainfall averages a pH of 4.5, with some mountain-hugging clouds as low as 2.0 in summer. A pH of this level is extremely acidic, toxic enough to damage both automobile paint and human skin surfaces. The Great Smokies also had the highest sulfur and nitrogen deposits of any monitored location in North America (Brown 2000).

There are two categories of ozone. One occurs naturally, the high stratospheric ozone that protects the Earth from harmful ultraviolet radiation. There has been much concern recently that this ozone layer is threatened because of atmospheric pollution resulting from the action of chlorofluorocarbons (CFCs). Increased levels of UV radiation, resulting from the depletion of stratospheric ozone, have been implicated in the decline of certain amphibians and in the high percentage of malformations in others. Fortunately, this does not seem to be a problem for the amphibians of the Southern Appalachians (Starnes, Kennedy, and Petranka 2000). However, the other form of ozone, surface-level (or tropospheric) ozone, results directly from human-caused pollution and may have serious consequences for amphibians. The ground-level ozone found within the park is formed when nitrous oxides mix with sunlight and natural hydrocarbons. This ozone is a powerful respiratory irritant and causes significant agricultural damage. In the park, ground ozone causes leaf damage, resulting in the injury and stunted growth of native plants. Ground-level ozone levels in the Great Smokies are twice as high as in Knoxville or Nashville and, in 1998, there were 44 days when the air was rated as "unhealthy" (an unhealthy rating is one in which ozone is measured at 85 parts per billion or greater). On one day, the ozone levels reached a record level of 135 ppb, a level at which humans experience significant respiratory distress. Suffice it to say, the air in the Smokies is no longer clean and healthful.

There are no studies that directly assess the effects of many atmospheric pollutants on amphibians, certainly within the Smokies. Indeed, many amphibians, because of their secretive habits, may be only indirectly impacted. For example, if atmospheric pollution affects vegetation cover and habitat structure, then it could indirectly affect the environmental conditions experienced by amphibians. On the other hand, low pH levels have rather deleterious effects. Acid rain leaches toxic chemicals and heavy metals, including aluminum and mercury, from soils and rock formations, such as the Anakeesta. The leachates, in turn, kill animals directly and otherwise influence distribution patterns. Acid soils are avoided by some terrestrial salamanders, as are acidic waters by aquatic species. Low pH is also known to inhibit the development of amphibian eggs and larvae, and to alter the sodium balance of adults (Wyman and Hawksley-Lescault 1987; Frisbie and Wyman 1991, 1992).

To my knowledge, there has been only one analysis conducted for toxic residues in the amphibians of the Great Smokies. No detectable residues were found in the tissues of either salamanders (Northern Slimy Salamander, Jordan's Salamander, Black-chinned Red Salamander) or frogs (American Toad, American Bullfrog, Northern Green Frog, Pickerel Frog) in an analysis for 29 pesticides and related compounds in tests conducted by biochemists at Virginia Tech for the USGS. However, sample sizes were small, and a more wide-ranging study needs to be conducted.

Nonindigenous Species

There are many nonindigenous species which inhabit the Great Smoky Mountains, particularly among the plants. Exotic plant diseases, such as the chestnut blight, and insects that kill native plants, such as balsam and hemlock adelgids, completely changed or are changing the entire vegetative community and structure. Other exotic diseases and insects that affect the forest of the Great Smokies include dogwood anthracnose, beech bark scale insects, mountain ash sawfly, and Dutch elm disease. Although these diseases and insects may have no direct effects on amphibians, altering the

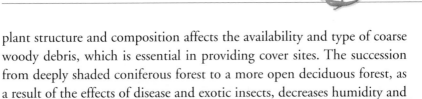

plant structure and composition affects the availability and type of coarse woody debris, which is essential in providing cover sites. The succession from deeply shaded coniferous forest to a more open deciduous forest, as a result of the effects of disease and exotic insects, decreases humidity and soil moisture, and increases the amount of sunshine and ground temperatures to which amphibians are exposed. Such changes could alter activity cycles, behavior, and prey communities and may change the environment so much that amphibian physiological tolerances, at least for sensitive high-elevation species, are exceeded.

Nothing promotes tourism in the Smokies more than trout fishing, and the desire to "improve" the sport led to certain decisions that radically affected amphibians and their habitats in the Great Smokies. The Southern Brook Trout was not providing sufficient sport for fishermen, so the National Park Service decided to introduce Northern Brook Trout and Rainbow Trout into certain park streams. During the 1930s, CCC crews built rearing pools at The Chimneys and Kephart Prong, and released between 300,000 and 400,000 15 cm (6 in) fish per year in park streams. The Izaak Walton League also maintained Rainbow Trout hatching ponds at Forney Creek, Elkmont, and Deep Creek. Although putting non-native fish into habitats where native species might be threatened has been against National Park Service policy since 1936, the restoration of native fish populations and NPS's own policy were ignored as late as 1971, when 20,700 Rainbow Trout were released at 27 locations in the Smokies. Such stocking not only endangered native trout populations, but probably reduced the population of aquatic salamanders since exotic trout are voracious amphibian predators. The introduction of nonindigenous trout into areas where they previously did not occur has contributed to the dramatic amphibian declines in the western United States (Resetarits 1997; Knapp and Matthews 2000; Matthews et al. 2001).

Stocking was not the only way the NPS facilitated trout fishing. To improve fishing in Indian Creek, it applied rotenone (a poison) to the creek over a 5-hour period in June 1957. Later, park rangers dumped rotenone into 24 km (14.6 mi) of Abrams Creek from Cades Cove to Lake Chilhowee. Forty-seven species of "rough" (that is, native) fish, including several endangered species, were killed to make way for trout. NPS records indicated there were "no survivors." In July of that year, 1,000 large Rainbow and 12,000 fingerling Northern Brook Trout were released in Indian

Creek, and 2,400 adult Rainbows and 30,800 fingerlings were released into Abrams Creek. Heralded as a success in 1957, this so-called restoration project is now known to have devastated the aquatic community. Where Hellbenders once thrived, they no longer occur. However, Mudpuppies have now recolonized Abrams Creek, since they could survive for a time in the deeper waters of then newly created Lake Chilhowee, as have streamside amphibians over the last 50 years. As the National Park Service currently assesses what will be necessary to restore the native fish community, consideration should be given to reintroducing Hellbenders into Abrams Creek, assuming an adequate prey base exists. Hopefully, the days are over when nonindigenous, predaceous fishes are deliberately introduced into eastern national parks.

In 1920, 100 European wild boars escaped from a tourist hunting preserve near Hooper Bald just outside the Smokies. Despite the damage that became quickly apparent, state wildlife officials even started a breeding program to aid hunters. The boars quickly spread to the Great Smokies, and today do a considerable amount of damage to the park's vegetation and wildlife communities (see, for example, Howe and Bratton 1976). They root up terrestrial surface litter, dig through the soil, overturn rocks and woody debris, destroy spring heads, and damage shallow wetlands used by amphibians. Of course, they eat every animal they find in the litter and soil. After a herd of hogs has passed through, the ground appears as if it has been bulldozed. During field surveys from 1998 to 2001, hog damage to amphibian habitat was observed in every portion of the Great Smokies.

Disease

In late June 1999, USGS researchers received reports of dead and dying amphibians from the margins of Gourley Pond in Cades Cove. Three dead or dying juvenile Pickerel Frogs *(Rana palustris)* were collected from the pond margin on June 29. On July 12, biologists returned to the site and collected three recently metamorphosed Spotted Salamanders *(Ambystoma maculatum)*. Although alive, the animals appeared to be in poor health and were retained for a brief period prior to their deaths. The six specimens were sent to the USGS National Wildlife Health Center in Madison, Wisconsin, for diagnosis. Viral hepatitis was found in two salamanders and one frog; viral pancreatitis in one salamander and one frog; fungal myositis in one salamander; and advanced autolysis in one salamander and two frogs. Viral cultures were not performed, so the results must be viewed as presumptive rather than confirmed. However, the virus most likely responsible for the infection was from a family of viruses termed iridoviruses. This group, particularly in the genus *Ranavirus,* is responsible for serious infectious disease in amphibians (Daszak et al. 1999; Chinchar 2002). The fungus, named *Ichthyophonus,* is primarily known to affect fish, but has been reported in Eastern Red-spotted Newts from West Virginia and in American Bullfrogs from Illinois.

In 2000, a second mortality event affecting a large number of salamander larvae and tadpoles was observed at Gourley Pond. A total of 25 affected animals of four species (Wood Frog, Spring Peeper, Marbled Salamander, Eastern Red-spotted Newt) were collected for analysis. Most animals had problems with their livers, spleen, pancreas, or kidneys. Necrotizing hepatitis, splenitis and mesonephritis due to iridovirus infection were found in 5 of 10 Wood Frog tadpoles; necrotizing hepatitis, glomerulitis, and pancreatitis were observed in one of three adult Eastern Red-spotted Newts; necrotizing hepatitis, pancreatitis, and mesonephritis were seen in all three Marbled Salamander larvae; and vesiculating stomatitis of the oral disc was observed in one Wood Frog tadpole, again due to the iridovirus infection. Positive iridovirus cultures were taken from one adult Eastern Red-spotted Newt and one Marbled Salamander larva. Oxyurid (pinworm) infections were seen in 11 Wood Frog tadpoles, trematodes in one newt, intestinal coccidiosis in one newt, and one tadpole (presumably a Spring Peeper) was normal.

Amphibian mortality recurred at Gourley Pond in 2001. A total of 15 animals of three species (Wood Frog, Southern Red-backed Salamander, two unidentified fish) were sent to the USGS National Wildlife Health Center. Pathology consistent with the presence of iridovirus in four of the 11 Wood Frog tadpoles was noted. No viruses were found in the terrestrial Southern Red-backed Salamander or in the fish. Four Wood Frog tadpoles also contained the fungus *Ichthyophonus,* although it is unlikely that the presence of this fungus would have caused the death of the tadpoles. *Ichthyophonus* was not reported from specimens collected in 2000.

The iridovirus outbreak at Gourley Pond did not affect all amphibians at the pond, although large numbers of larval amphibians died. Iridovirus outbreaks were not recorded at any other site within the park, including the many wetland breeding sites elsewhere in Cades Cove. The cause and long-term effects of the iridovirus infection at Gourley Pond are unknown. Amphibian iridoviruses may survive long periods at the bottom of ponds, particularly during winter. Serious outbreaks of iridovirus are known or suspected to be involved in at least some of the declines reported for amphibians elsewhere in North America and in Europe (Daszak et al. 1999; Chinchar 2002). These findings suggest that additional long-term monitoring and research should be directed at understanding the health status of amphibians breeding at Gourley Pond and perhaps elsewhere in Cades Cove.

In addition to the diseased specimens found at Gourley Pond, one large female larval Spring Salamander from the Cane Creek drainage was found in 2000 with a nodule on the side of her head. The nodule was the result of a fungus-like organism, tentatively referred to *Dermocystidium*. Fortunately, no malformations have ever been observed in the amphibians of Great Smoky Mountains National Park.

Conserving Amphibians in the Smokies

Many amphibian species are in serious decline. In parts of the world, such as the wet tropics of Australia and Central America and the high mountain regions of western North America, entire amphibian communities have nearly disappeared. The causes of declines are diverse, and no one cause is likely responsible for the majority of declines. Diseases (both viral and fungal), ultraviolet radiation, toxicants (pesticides, herbicides, industrial chemicals, endocrine-mimicking compounds), acid rain, weakened immune systems, and over collecting all have been implicated to one degree or another. These factors may not act alone, but in synergy. For example, exposure to a pesticide might not kill an animal outright, but instead weaken its immune system, making it more susceptible to disease, especially in times of extreme cold or drought when the animal is naturally stressed. Sublethal effects might also slow swimming speeds or decrease feeding ability and growth, thus making an animal more prone to predation or to pond desiccation. However, the greatest threat to amphibians remains the destruction or alteration of habitats.

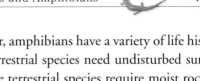

As discussed in the first chapter, amphibians have a variety of life histories and habitat requirements. Terrestrial species need undisturbed surface and underground retreats; some terrestrial species require moist rock crevices or exfoliating bark, whereas others reside in talus slopes. Many aquatic species require clean, cold, flowing streams and rivers in the Southern Appalachians, whereas others require permanent ponds. Those species with a truly biphasic life cycle, whereby they reproduce in water but reside on land most of their lives, are most difficult to conserve (Semlitsch 2000). Not only must breeding sites be maintained, but also the terrestrial retreat and overwintering sites. Inasmuch as terrestrial habitats may be located hundreds of meters from a breeding pond, small buffer zones around wetlands or narrow riparian corridors along streams are not sufficient to protect an amphibian community (Semlitsch 1998). In addition, migratory routes need to be maintained between breeding and terrestrial sites, and prey populations must not be seriously impacted by environmental stressors.

Significant outright habitat destruction is no longer a serious problem within Great Smoky Mountains National Park. Relatively small amounts of habitat occasionally are disturbed in connection with the maintenance of visitor facilities and road improvements, but the massive "development" envisioned by early park promoters, and the wetland drainage schemes of the old Soil Conservation Service, are apparently not viable options in the twenty-first century. Care must be taken in any consideration of facility and road expansion, especially within Cades Cove, however. Significant habitat loss could occur if Lakeshore Drive is ever completed, especially since fragile habitats, important to amphibians, are located in the lower reaches of Hazel, Eagle, and Twentymile Creeks. Likewise, road-widening projects around the periphery of the park should be approved only after surveys have demonstrated that no impacts will occur to sensitive species, such as Hellbenders and the Junaluska Salamander.

Today, the most serious dangers to the amphibians of the Great Smoky Mountains come from external threats: disease, airborne pollution, and nonindigenous species. Disease, in the form of iridovirus, has been detected in several amphibian species at one locality in Cades Cove. Iridoviruses have long been known to cause amphibian mortality, but the effects tend to be localized. Not all species or even individuals become infected, and the long-term impacts of iridovirus outbreaks on amphibian populations are unknown. Fortunately, the extremely serious fungal disease, chytridiomycosis, has not yet been discovered within the park. If it ever is, the park's amphibians will face a serious threat to their survival. For the moment, the best way to prevent the spread of disease is to limit exposure to impacted areas. Boots and sampling equipment used during visits to Gourley Pond should be treated with a bleach solution immediately upon exiting the pond. Rubber, but not latex (which is toxic to amphibians), gloves should be worn when collecting or sampling pond water. Persons visiting this pond for any reason should ensure that biosafety procedures are incorporated as part of their visit to the pond's basin. Although the disease is not known to affect humans, visitors should never pick up dead or dying amphibians. The presence of dead amphibians should be reported to the Twin Creeks Natural Resource Center of the National Park Service at Great Smoky Mountains National Park.

The only way to lessen the threats to Great Smokies amphibians from airborne pollution, acid rain, and ozone is to call for the strict enforcement of existing pollution laws and the promulgation of standards, where necessary, to lower the levels of these hazards emitted into the atmosphere. Clean-burning fuels and stringent emission standards would help all biota, including humans and amphibians, both within and outside the park. The limitation of vehicular traffic in Cades Cove, currently under review, might also help to reduce pollution levels and to curb runoff from vehicles, such as motor oils, that are known to affect amphibians adversely. Still, the effects of pollutants originating directly from within the park are probably localized, and pollution levels are low. It is the external threats, often originating far from the park's boundaries, that potentially have the most serious consequences.

Although nonindigenous plants and insects probably have had little direct effect on the amphibians of Great Smoky Mountains National Park, they certainly have affected the communities in which amphibians live. The loss of the American Chestnut forest undoubtedly altered the structure of amphibian habitats, especially through changes in the amount of available hiding places and by the extinction of invertebrates that depended upon the chestnut. As mentioned in the second chapter, the loss of the American Chestnut also may have affected the distribution of the Green

Salamander in the Great Smokies. Biologists will never know for sure to what extent the complete change from a chestnut forest to a forest dominated by a variety of oaks influenced amphibian populations, since baseline data do not exist.

In a similar context, the loss of Fraser Firs at higher elevations in the park today may have no immediate effects, or the change from a coniferous forest to one dominated by hardwoods might completely alter habitat structure. If the habitat structure changes, the amounts of humidity, moisture, and coarse woody debris could also change, thus influencing endemic salamander communities. Other imported insects affect individual tree species, such as the hemlock woolly adelgid, which feeds on Eastern Hemlock trees, and the beech bark scale insect. Unfortunately, not much can be done at present to stem the effects of the balsam woolly adelgid and perhaps most of the other introduced pests, so biologists are forced to monitor the situation and record what happens. With regard to the balsam woolly adelgid, the accidental introduction of this tiny insect resulted in billions of dollars in damages, lost income, and ecosystem disruption. Tighter controls need to be in place to prevent such damage from reoccurring again and again.

The National Park Service no longer introduces non-native trout into the park, although efforts to re-establish the native book trout continue. In 1999, NPS proposed to treat 4 km of Sams Creek and a number of additional streams within the Great Smokies with antimycin to kill non-native Northern Brook and Rainbow Trout. According to the draft Environmental Assessment, this action would likely have no effect on larval and adult salamanders and frogs, based on field trials where both larval and adult salamanders survived for 8 hours at 8 ppb antimycin (National Park Service 1999). The introduction of any poison into a stream with amphibian populations should be monitored closely, especially within a National Park.

The most deleterious nonindigenous species within Great Smoky Mountains National Park continues to be the European wild boar. This species is extremely destructive of plant and animal life, surface leaf litter, and small seeps and wetlands. Efforts to eradicate this pest should be encouraged, not only to benefit amphibians, but for the benefit of the entire mountain ecosystem.

Searching for Amphibians

As residents of a national park, the amphibians of the Great Smoky Mountains are strictly protected. No frogs or salamanders should ever be collected or removed from the park, unless written permission has been obtained from the National Park Service. Visitors should enjoy amphibians in their native habitats, and leave them to carry out their roles as predators and prey in the natural world. Most amphibians make poor pets, and taking them from their homes deprives future generations of their presence and contribution to the gene pool. Likewise, amphibians from other environments or locations should never be released into the park. Such individuals may carry disease or may otherwise deleteriously affect native species. Take nothing but pictures; leave nothing but footprints.

Amphibians often are observed as they wander over the forest floor, hide under leaf litter and rocks, crawl up wet rock faces, or call from streams, ponds, and other small wetlands. If rocks or other surface objects are moved while searching for these often secretive animals, they should always be replaced in the exact position where found. Avoid habitat disturbance as much as possible. Never damage the park's environments, for example, by destroying rock crevices or stripping tree bark, even on fallen trees. Penalties for violating park laws are severe, and they are enforced.

There are several good ways to view the park's amphibians. Several species of dusky salamanders are easily observed during the day as they sit and wait for prey along streams and rock faces; a visitor might only glimpse a head with two, large, beady eyes poking out from a wet crevice, yet still be able to identify it as a Seal Salamander *(Desmognathus monticola)*. Observe salamander larvae crawling along in shallow river or stream waters. Listen for calling frogs and attempt to find the caller—it is often quite a challenge. Take a short walk on a rainy or misty spring or summer night and search pond margins, rock faces, surface litter, trees, seeps, and streams. Larval and adult amphibians that are hard to find during the day are much more likely to be seen at night, even when it is not raining.

Take field notes of the species seen, unusual color patterns, habitats where the animals were found, and the environmental conditions under which amphibians were observed. Recording such variables will offer a much greater appreciation of the animals than by simply observing them and will often help to identify the individual. A camera is a valuable asset, and the new digital cameras offer an immediate ability to compare photographs taken in the field with illustrations in guidebooks. These cameras are very effective for amphibian "close-ups," and they avoid the focusing and depth-of-field problems often experienced using standard SLR cameras. A good tape recorder will help in the identification of frog calls, especially since many frog calls now can be accessed through the Internet. For example:

http://www.cars.usgs.gov/herps/Frogs_and_Toads /frogs_and_toads.html

and

http://www.state.tn.us/twra/frogs.html)

The more visitors know about an animal observed in the park, the more they should come to appreciate that animal's place in the environment, which will result in a greater appreciation of the importance of preserving habitats even outside the park.

Identification

Most visitors to Great Smoky Mountains National Park, even those without advanced biological training, should be able to identify the majority of the amphibians they observe by using a combination of the color photographs, species descriptions, and identification/life history tables found in this chapter and in the following species accounts. Some individual animals, however, may be impossible to identify with certainty, even by specialists. Larvae, especially small salamander larvae and tadpoles, often cannot be distinguished without resorting to microscopic examination, and adult salamanders, especially the duskies *(Desmognathus)*, are notoriously variable with overlapping phenotypic and genotypic characters. Field biologists have found it increasingly difficult to place some individual animals into a species category because of the range of genetic and color variation observed in natural populations. Many times during our surveys we found ourselves in concentrated discussion about the identification of a particular animal, and sometimes we recorded an animal as unknown in our field notes.

One of the best ways to identify salamander and frog larvae, in addition to color and morphology, is to examine their habitats and the times of year they are found. This can most easily be done through a comparative table. In tables 3 and 4, I provide a list of morphological and life history characteristics that should aid a visitor in determining which species is being examined. These data then can be used in conjunction with the information in the species accounts to accurately identify specimens in hand.

Table Three

Species	Egg Deposition	Hatching	Larval Period	Hatching Size	Size and time of Metamorphosis
Spotted Salamander	Jan to late Mar (mountains: late Feb to early Mar); 4–7 weeks incubation	Apr–May	2–4 mos.	12–17 mm TL	29–32 mm SVL; 43–60 mm TL (to 75 mm TLif overwinter); mid-June to Aug
Marbled Salamander	Oct–Nov (in pond by Sept); 9–15 days incubation, but must be flooded 1–2 days	winter	5–7 mos	10–14 mm TL	~33 mm SVL; 49–58 mm TL; late Mar– mid-June
Mole Salamander	Sept to Mar (winter)	winter to early spring	3–4 mos, but variable	~10 mm SVL	32–50 mm SVL; May to Sept
Spotted Dusky Salamander	early May to early July, perhaps to mid-Aug; 45–60 days incubation	July to early fall	<1 yr	8–12 mm SVL; 12–20 mm TL	9–12 mm SVL, to 20 mm SVL; July to early fall?
Imitator Salamander	late spring to early summer?	July to early fall?	<1yr		10–12 mm SVL; July to early fall
Shovel-nosed Salamander	late spring to early summer?; 10–12 weeks incubation	mid-Aug to mid-Sept	3 yrs (10–20 mos)	11 mm SVL	26–38 mm SVL; May to Oct.
Seal Salamander	mid-June to mid-Aug; 2 mo incubation	early summer to fall; Sept	10–11 mos	11–12 mm SVL	June–July
Ocoee Salamander	July to early Aug, to Sept; 52–74 days incubation	Aug to late Sept	9–10 mos	13–18 mm TL	11–15 mm SVL; May to June
Black-bellied Salamander	May to June	July to Sept	3–4 yrs	11–16 mm SVL	35–42, to 54 mm SVL; midsummer
Santeetlah Salamander	early May to early July, perhaps to mid-Aug; 45–60 days incubation	July to early fall	<1 yr	8–12 mm SVL; 12–20 mm TL	9–12 mm SVL,to 20 mm SVL; July to early fall?
Three-lined Salamander	winter	early to mid-Mar?	3.5–5.5 mos. (< 1 yr), but may overwinter	11–12.5 mm SVL	22–27 mm SVL, to 32 mm SVL; June to Aug
Junaluska Salamander	at least by mid-May	early June?	1–2 years	7–9 mm SVL; 11–13 mm TL	34–42 mm SVL; mid-May o Aug
Long-tailed Salamander	late autumn to early spring	Nov–Mar after 4–12 weeks	normally <1 yr (4–7 mo)	18–21 mm SVL; 40 mm TL	23–28 mm SVL; >50 mm TL if overwintering;mid-June–July
Cave Salamander	Sept to Feb		6–18 months; most 12–15	9–12 mm SVL; to 17.5 mm TL	31–37 mm SVL; to 70 mm TL; spring
Blue Ridge Two-lined Salamander	Feb to May; 4–10 weeks incubation	May to Aug	1–2 years	7–9 mm SVL; 11–14 mm TL	18–19 mm SVL in 1 yr, to 32 mm SVL in 2 yr; Apr to July
Spring Salamander	summer	late summer to autumn	to 4 years	18–22 mm TL	55–65 mm SVL,to 70 mm high elevation; late June to August
Four-toed Salamander	Feb to May	May–June?	21 to 61 days		11–15 mm SVL; 17–25 mm TL; July?
Mud Salamander	autumn to early winter	winter	15–17 mos to 29–30 mos	<13 mm SVL	35–44 mm SVL; mid-May to Sept.
Black-chinned Red Salamander	autumn to early winter; 3 mo incubation	mid-Dec to mid-Feb	1.5 tFo 3.5 yrs (27–31 mos)	11–14 mm TL	34–46 mm SVL; 62–86 mm TL; May to July

Table Three

Spots on dorsum	Dorsal Pattern	Belly Pattern	Tail Attributes	Notes
	dull olive green, no conspicuous markings	white or light	tail fin lightly mottled or finely stippled; dark at tip	Breeding occurs in 2–3 bouts following rain; pond type larvae
	blackish, drab; older larvae have mottling on body	throat stippled; scattered melanophores on lateral sides	dorsal fin extends	Pond type larvae; series of ventrolateral light spots forming a line below limb insertions
	black and yellow blotches along midline of back	dark band on midline (poor in some specimens)	almost to front limbs yellow and black on tail fin	Pond type larvae; variable life histories with regard to timing of events
5–8 pairs of even or alternating spots or blotches	sides with dorso-lateral stripe; dorsum variable		spot pattern continues on tail	In older larvae, spots or blotches may fuse
		dark		cheek patches may be present on larvae
2 rows light spots	dark, conspicuous light flecks on sides		spatulate	more slender, w/ longer legs than DQ
4–5 Fpairs light dorsal spots between limbs				
4–6 pairs of alternating light spots on dorsum				round snouts
6–8 pairs light spots between limbs	light brown			much larger than all other Desmogs; lots of yolk 1–2 mos after hatching
4–5 pairs of even or alternating spots or blotches				
no paired light spots	cream; uniformlystippled; then dark broad dorsolateral stripe; narrow mid-dorsal stripe	immaculate	dorsal fin does not extend forward of rear legs	stream type
	deep olive green to brown	no iridophores		dense, well-defined cheek patches; lower margin of dark pigmentation straight
	cream colored; then uniformly dark, similar to adults; no paired spots	immaculate		stream type; tail fin stops near insertion of rear limbs; reddish gills; square snouts
	sparse pigmentation with 3 longitudinal series of spots on the side			
6–9 pairs light dorsolateral	dusky	light with iridophores		
	light yellow brown to gray with fine flecking			long truncated snouts with small eyes
	nondescript, yellow brown; dorsal fin extends to head			pond type larvae; joint nesting occurs; brooding
	light brown; older w/widely scattered spots	immaculate		stream type; overwintering occurs; larvae can be very large
	light brown; weakly mottled or streaked	dull white		stream type; no black chins or dorsal spots

Salamanders

All salamanders in the Great Smoky Mountains have four limbs with four *(Necturus, Hemidactylium)* or five (all others) toes on each hind foot. They all have tails, lack dry scales covering the body (lizards have dry scales), and have skins that are moist or wet to the touch. The skins of a few species, such as Jordan's Salamander, are sticky because of glandular secretions, but only the Hellbender and Common Mudpuppy are truly slimy. Biologists take two standard measurements with regard to length. The total length (TL) is the length of an animal from the tip of the snout to the tip of the tail. Because some salamanders lose their tails (or parts thereof), another common measurement recorded is the snout-vent length (SVL). SVL is measured from the tip of the snout to the posterior portion of the vent (the opening of the cloaca, the common receptacle for the digestive, excretory, and reproductive tracts). All scientific measurements are recorded in metric units, usually millimeters, although both metric and English measurements are frequently mentioned in the species accounts.

Salamander larvae sometimes are divided into two general groups, depending on morphology and the type of wetland in which they develop. The pond form is stout bodied, with long filamentous gills and a wide dorsal fin which extends well onto the body. Mole salamanders *(Ambystoma),* for example, have this type of larva. Pond larvae develop in still water, and use the extra surface area of the body and fin as aids in swimming. Stream larvae are slimmer than pond larvae, with more streamlined bodies, shorter gills, and a narrower tail fin that does not extend onto the body. These larvae usually live in swift flowing water, where extra surface area on the body would be a distinct disadvantage. The genera *Eurycea* and *Pseudotriton* have this type of larva.

There are a number of useful characters which can be used to identify salamanders to genus or family. A few illustrative examples are provided below, but more detailed comparisons are found in the species accounts under the heading similar species.

Desmognathus: all dusky salamanders have a light line which extends from the back of the eye to the angle of the jaw. The duskies also have well-developed muscles on the sides of their heads. They need these muscles to raise the upper jaw in order to open their mouths, since the lower

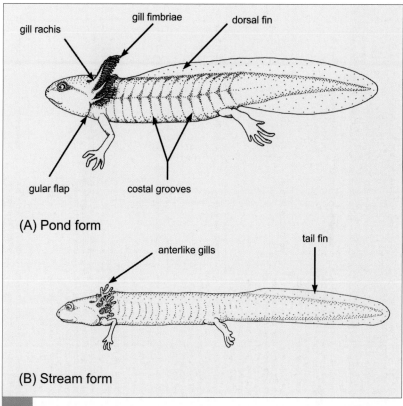

Body morphology of a salamander larva. (A) Pond form. (B) Stream form.

jaw is fused to the skull. This is exactly the opposite of the way most animals, including humans, open their jaws.

Gyrinophilus versus *Pseudotriton:* although these colorful salamanders are superficially similar in appearance, Spring Salamanders *(Gyrinophilus)* have a canthus rostralis, a large white line bordered by black lines that runs from in front of each eye to the nostril. Salamanders of the genus *Pseudotriton* do not have this line. Spring Salamanders use the canthus rostralis as a "gunsight" to zero in on prey.

Plethodontidae versus all other salamander families: all lungless salamanders have a nasolabial groove that goes from each nostril to the upper jaw. The nasolabial groove transmits chemicals to the salamander from the substrate, and no other salamander families have this groove.

Table Four

Species	Eggs	Tadpole Description	Breeding Times	Larval Period	Metamorph Size
Northern Cricket Frog	eggs deposited singly; 1 gelatinous envelope, >2.3 mm in diameter; deposited in shallow water among stems of grass or on bottom; 250 eggs per complement	a medium-sized light to medium gray tadpole; throat light; tail musculature mottled or reticulated; usually a very distinctive "black flag" on the tail tip; tail long and narrow; anus dextral (to the right); oral disk emarginate; most 30–36 mm TL, rarely to 46 mm.	breeds Apr to June, possibly into July	35–70 days, based on Acris crepitans blanchardi	10–15 mm
American Toad	eggs in strings with gelatinous casings; 2 envelopes present; strings long, to 60 m; 15–17 eggs per 25 mm; 4,000–12,000 eggs on bottom of quiet pools	body round or oval in dorsal view; eyes dorsal (looks cross-eyed); nostrils large; color dark brown to black; dorsal portion of the body unicolored; venter with aggregate silvery or copper spots; snout sloping in lateral view; tail musculature distinctly bicolored; anus medial (in the center); spiracle is distinctly on left side of body	breeds spring (Mar–Apr)	50–65 days	7–12 mm
Fowler'sToad	eggs in strings with gelatinous casings; 1 envelope present and <5 mm in diameter; strings 2.4–3 m with 17–25 eggs per 25 mm; 5,000–10,000 eggs; in tangled mass around vegetation	body round or oval in dorsal view; eyes dorsal (looks cross-eyed); nostrils large; color dark; dorsal portion of body slightly mottled; snout rounded in lateral view; tail musculature often not distinctly bicolored; anus medial (in the center); spiracle is distinctly on left side of body	breeds Apr to July	40–60 days	7.5–11.5 mm
Eastern Narrow-mouthed Toad	eggs in small surface film that has a mosaic structure; envelope a truncated sphere; mass round or square; 10–150 eggs per mass; in any depression with water, but not deep pools	a small jet black tadpole with lateral white to pink stripes on posterior portion of body extending to the tail musculature. Viewed from the side, the head comes to a point; body round in dorsal view; eyes wide set and lateral; anus median; jaws do not have keratinized sheaths, and the oral disc and labial teeth are absent	breeds mid-May to mid-Aug	20–70 days	8.5–12 mm
Cope's Gray Treefrog	eggs in small surface film, but envelope not in truncated sphere; no mosaic structure; 5–40 eggs per mass; in shallow ponds attached loosely to vegetation, or free. Air bubbles present.	small to medium-sized grayish tadpole with a high dorsal tail fin; dorsal tail fin height equal to or greater than musculature height; tail long, with black blotches; background color of mature tail orange to scarlet; throat rarely pigmented; dorsal fin never extends anterior to midway between the spiracle and eye; anus dextral (to the right); oral disk not emarginate	breeds April to June, but calls occasionally heard at other times of the year	45–65 days	13–20 mm
Spring Peeper	eggs deposited singly in shallow water near bottom among vegetation; one gelatinous envelope.	a small-sized deep-bodied tadpole with a medium-sized tail; tail musculature mottled; fins clear or with blotches; no dots on body; snout square when viewed dorsally; anus dextral (to the right); oral disk not emarginate	breeds late winter to early spring (Feb to Apr); calls occasionally heard at other times of the year	90–100 days	9–14 mm
Upland Chorus Frog	egg mass in lump, but loose irregular cluster; 1 envelope, 3.6–4.0 mm; deposited in marshy areas and pools in matted vegetation	small olive to black tadpole with a bronze belly; tail medium; anus dextral (to the right); oral disk not emarginate; tadpoles develop rapidly	breeds February to April.	50–60 days	8–12 mm

Species	Eggs	Tadpole	Breeding	Larval Period	Size
American Bullfrog	eggs in large surface film in form of a disc; 10,000–12,000 eggs per disc; deposited among water plants or brush; 1 gelatinous envelope	large olive to grayish green tadpole with small widely spaced small spots (dots) covering the body and tail; venter straw; eyes bronze; body oval and round in dorsal view; eyes dorsal or dorsolateral; nostrils small compared with eyes; lower jaw wide; anus dextral (to the right); oral disk emarginate	breeds late spring throughout the summer. Calls may be heard at other times of the year	1–2 yrs	31–59 mm
Northern Green Frog	eggs in surface film: mass <0.09 sq m; 1,000–5,000 per mass; attached to vegetation or free; 2 gelatinous envelopes	large (but not deep bodied) olive green tadpole with large dark spots, generally with a white throat; belly deep cream without iridescence; body oval and round in dorsal view; eyes dorsal or dorsolateral; nostrils small compared with eyes; tail green mottled with brown; lower jaw wide; anus dextral (to the right); oral disk emarginate	breeds late April to late July or even early Aug. Calls may be heard at other times of the year	to 1 yr	23–38 mm
Pickerel Frog	eggs in firm regular cluster; brown above and yellow below; mass a sphere 38–100 mm in diameter; 2 envelopes present; 2,000–4,000 eggs; mass deposited 75–100 mm to 91 cm under water; attached to debris and vegetation	large, full, deep-bodied tadpole; olive green shading through yellow on sides; venter cream, back marked with fine black and yellow spots; belly with blotches of white; venter iridescent, viscera visible; tail very dark, black blotches can aggregate to purple-black; body oval and round in dorsal view; eyes dorsal or dorsolateral; nostrils small compared with eyes; lower jaw narrow; anus dextral (to the right); oral disk emarginate	breeds late winter to spring (mid-Mar to Apr)	70–80 days	19–27 mm
Northern Leopard Frog	mass a firm regular cluster; 3,500–6,500 eggs close together in mass; 2 envelopes present; outer envelope 5 mm; eggs black above and white below; deposited near surface, usually attached to grasses and vegetation, sometimes free	large, deep-bodied tadpole; dorsally dark brown, covered with small gold spots; belly deep cream, with bronze iridescence; viscera visible; throat translucent and more extensive than Pickerel Frog; similar in appearance to Green Frog, but darker; body oval and round in dorsal view; eyes dorsal or dorsolateral; nostrils small compared with eyes; lower jaw narrow; anus dextral (to the right); oral disk emarginate	breeds probably early Mar to early May	60–80 days	18–31 mm
Wood Frog	eggs in firm regular cluster; black above and white below; mass a sphere 38–100 mm in diameter; 2 envelopes present; 2,000–4,000 eggs; mass deposited 75–100 mm to 91 cm under water; attached to debris and vegetation	medium-sized tadpole with usually very dark to gray coloration, and with a faint light stripe of cream, white or gold along the upper jaw (like a mustache); venter cream with belly slightly pigmented at sides; body oval and round in dorsal view; eyes dorsal or dorsolateral; nostrils small compared with eyes; anus dextral (to the right); oral disk emarginate; tail quite long; dorsal crest high extending on to body	breeds winter and early spring (mid-Dec to Mar)	45–85 days	16–18 mm
Eastern Spadefoot	eggs in loose irregular cylinder or band; mass 25–75 mm wide and 25–305 mm long; deposited on stems of plants/grass; 1 gelatinous envelope; 200 per packet	a small, dark tadpole, bronze to brown with close set tiny orange spots; body round or oval in dorsal view; eyes close set and dorsal, iris black; head wide relative to body width; tail short, with tip blunt and rounded; anus medial (in the center); spiracle is ventrolateral. Often found in "schools" of hundreds of tadpoles	breeds: only heard calling once (July 12, 1999). Probably any time from Mar to Oct	14–60 days	8.5–12 mm

Frogs

Like most salamanders, frogs have four legs with four toes on the front limbs and five toes on the rear limbs. The hind limbs are much larger than the front limbs, and are used to propel the body when walking, hopping, or jumping. Frogs are measured in TL, that is, from the tip of the snout to the end of the body between the hind limbs. Of course, there are many other measurements that specialists make, such as the length of the various sections of the hind limb, but these are not important for identifying the frogs of the Great Smoky Mountains.

Tadpoles are quite complex morphologically. As with salamanders, there are two general body types of tadpoles, the pond type and the stream type. Pond-type tadpoles have deeper bodies and higher tail fins than stream-type tadpoles. Structures important in the identification of tadpoles are labeled in the figure below. The oral disk consists of the mouth parts; the narial aperture is the opening to the nostrils; the spiracle is the opening from the gills (water is taken in through the mouth, passes over the gills, and is expelled via the spiracle); the anus is the opening from the digestive tract. The TL consists of the body length (BL) and tail length (TAL). Sometimes additional morphological measurements are taken, such as the maximum width of the tail musculature (TMH) or the maximum tail depth (MTH). The location and size of these characters, or their ratios in relation to one another, may be useful in identifying what otherwise appears to be just another drab, olive-green, or black tadpole.

The tadpoles of different species of frogs often appear extremely similar to one another, but their mouthparts readily separate them. Although most visitors are unlikely to look into a tadpole's mouth, biologists and naturalists need to be able to examine mouthparts to determine which species is being examined. For this reason, and because such information on Southern Appalachian frogs is not readily available elsewhere, I have included a diagram of tadpole mouthparts with each frog's species account. The nomenclature follows Altig and McDiarmid (1999) (see figure below). The location, number, and degree of separation among labial teeth and papillae are important characters for identifying tadpoles. Examining tadpole oral disks (sometimes incorrectly termed teeth) also gives biologists an

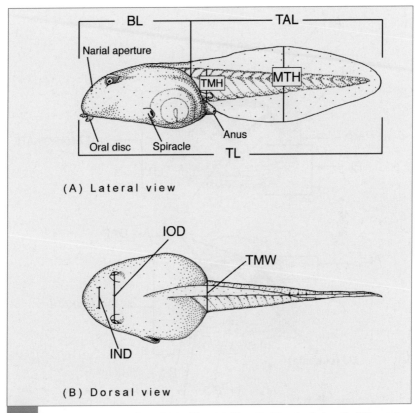

Body morphology of a tadpole. TL = total length; BL = body length; TAL = tail length; TMH = the maximum height of the tail musculature; MTH = maximum height of the tail, including both the tail fins and tail musculature; IND = distance between the narial aperatures (internarial distance); IOD = distance between the eyes (the interorbital distance); TMW = maximum width of the tail.

opportunity to check the health of the tadpole. For example, the cornified jaw sheaths drop out when the tadpole is exposed to some toxic compounds and to the dangerous disease chytridiomycosis.

As with salamanders, there are certain useful defining characteristics that help to identify certain superficially similar animals. Some of these are listed below (also see the heading similar species in the species accounts).

Bufonidae versus Pelobatidae: frogs in both of these families are terrestrial. However, the true toads *(Bufo)* are dry-skinned and "warty," and

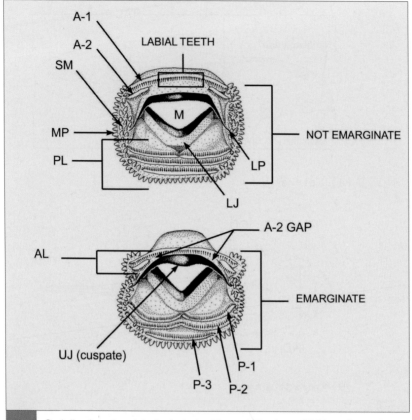

Oral disc (mouthparts) of a tadpole. AL = anterior (upper) labium; A-1 and A-2 = first and second anterior (upper) tooth rows; A-2 GAP = medial gap in second anterior tooth row; LJ = lower jaw sheath; LP = lateral process of upper jaw sheath; M = mouth; MP = marginal papilla; OD = oral disc; PL = posterior (lower) labium; P-1, P-2, P-3 = first, second, and third posterior (lower) tooth rows; SM = submarginal papilla; UJ = upper jaw sheath. Terminology is that of Altig and McDiarmid (1999:35).

have prominent cranial crests and parotoid glands. The spadefoots *(Scaphiopus)* are smooth-skinned, lack cranial crests and parotoids, and have a sharp digging spade on their hind feet.

Hylidae versus other frog families: all hylid frogs *(Acris, Pseudacris, Hyla)* in the Great Smokies have slightly to completely expanded toepads, but only in the treefrogs *(Hyla)* are they greatly expanded for climbing; the other hylids are mostly ground dwelling.

Pickerel Frogs versus Northern Leopard Frogs: these very similar frogs are both green and spotted. In Pickerel Frogs, the spots are squarish, paired and of nearly equal size, whereas in Northern Leopard Frogs they are smaller, rounded, and more randomly scattered on the frog's back.

Accounts of Species

As stated in the preface, much of the impetus for this book resulted from four years of intensive amphibian surveys in every part of Great Smoky Mountains National Park. More than 500 sites were sampled (map 1), and more than 10,000 amphibians were examined. The species accounts include information on etymology, identification of adults, larvae, and eggs, similar species and how to differentiate them, taxonomic problems, distribution both within the park and elsewhere in North America, life history, abundance and status, and remarks on interesting aspects of the biology of

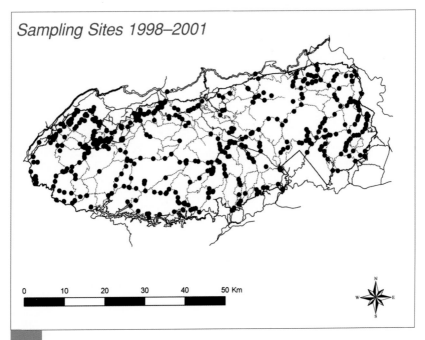

Location of U.S. Geological Survey sampling sites from 1998 to 2001.

the species. Information on 44 amphibians is included, including those species no longer thought present (for example, the Green Salamander) or which were reported historically from the park, but whose actual occurrence may be doubtful (the Northern Cricket Frog). Distribution maps, color photographs of amphibians from the park, and original color illustrations accompany each account. The maps provide data on the distribution of species as determined from USGS surveys from 1998 to 2001. Important historical locations, as determined from literature records or museum specimens, are included as squares, especially for rare species and to delimit historical range within the park. Whereas I have checked many museum specimens, I have not included all these data on the maps since many specimens in museums have not been properly allocated to species based on recently available genetic data, and most specimens are not georeferenced.

Accounts of Species

Salamanders

Spotted Salamander
Ambystoma maculatum

Etymology

Ambystoma: from the phrase *ana stoma buein,* to cram into the mouth; *maculatum:* from the Latin *maculatus,* meaning spotted. Thus, the scientific name refers to a spotted salamander with a voracious appetite, which is quite appropriate for this species.

Spotted Salamander *(Ambystoma maculatum)* adult, Gum Swamp in Cades Cove.

Identification

Adults. Adult Spotted Salamanders are robust animals, with large yellow to red to orange spots on a dark gray to black body. Spotting extends onto the tail. Occasionally, the spots are reduced or nearly absent, but this condition is rare in the Smokies. The bellies are light to dark gray. Mature females are larger than males, and sexually mature males in the breeding season are easily identified by their swollen vents. Adults measure 15–25 cm (6–10 in) TL.

Spotted Salamander recent metamorphs, Methodist Church Pond in Cades Cove (September 2000).

Larvae. The larvae are darkly colored (greenish to gray on the dorsum and sides), although the bellies are light colored and unpigmented.

95

Spotted Salamander larva, Methodist Church Pond in Cades Cove (July 10, 2000).

Spotted Salamander
(Ambystoma maculatum).

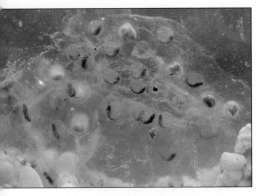

Spotted Salamander eggs, showing larval development and symbiotic green alga, Big Spring Cove (April 2000).

The abdominal vein is not generally visible. The tail fins are narrow in comparison to other *Ambystoma* larvae in the park, and the tail is relatively short.

Eggs. Up to 250 eggs are deposited within a clear or white jelly mass, and females deposit 2 to 4 egg masses per breeding season. The eggs are attached to sticks and aquatic vegetation, or they are placed directly on the pond bottom. Egg masses are quite firm; this ensures protection from predators, a defense against desiccation, and adequate spacing for oxygenation during development. Occasionally, egg masses are colonized by the green alga *Oophilia amblystomatis,* which helps to increase the oxygen supply. Both Wood Frog tadpoles and Eastern Red-spotted Newts will eat the eggs and newly hatched larvae.

Similar Species. Adults of this species cannot be confused with any other salamander currently known from the park. Recently transformed individuals are distinctive in their body morphology (they have a robust body with stocky legs and a relatively short tail), but their poorly developed spots may lead to confusion with other very recently transformed salamanders of this genus, the Mole and Marbled Salamanders. However, these other species usually have either transformed before the spotted (the marbled), or are very rare within the park (the mole). Robust animals with large orange to red spots on a dark body undoubtedly are Spotted Salamanders.

Taxonomic Comments. No subspecies are recognized.

Distribution

Spotted Salamanders are found in deciduous forest habitats throughout eastern North America, from the Saint Lawrence River and southern Ontario south to the Mississippi Gulf Coast, and west to eastern Texas and Oklahoma. They are absent from the prairie and grassland communities of the Midwest.

Spotted Salamander egg mass, showing predation by Wood Frog *(Rana sylvatica)* larvae. Ditch along road to Tremont (April 2000).

In the Great Smokies, they are found in ponds, woodland pools, and roadside ditches at lower elevations in the Little River–Abrams Creek drainages on the Tennessee side, particularly in Cades Cove. Breeding localities include ponds (Gum Swamp, Gourley Pond, Methodist Church Pond, The Sinks), shallow woodland pools (in the Cane

Spotted Salamander exposed egg mass, still containing viable developing eggs. Big Spring Cove (April 1999).

Creek, Beard Cane Creek/Shell Branch, and Meadow Branch drainages), and ditches (e.g., along the road to Tremont and the Cades Cove Loop Road). Huheey and Stupka (1967) mention its occurrence at 670 m (2,200 ft) at Elkmont (collected in 1935 and 1938) and in the Sugarlands area, but surveys from 1998 to 2001 could not relocate any sites in these regions. The park's research collection also contains one specimen collected in Whiteoak Sink in 1958. It seems probable that this species breeds in any woodland pool at low elevations on the Tennessee side of the park.

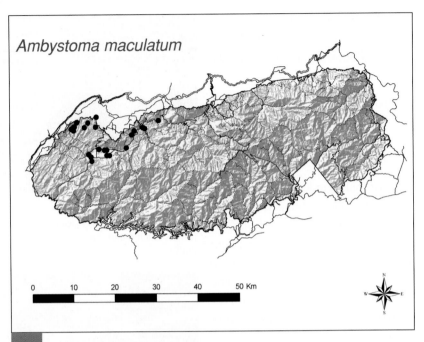

Ambystoma maculatum

| 0 | 10 | 20 | 30 | 40 | 50 Km |

Distribution of the Spotted Salamander *(Ambystoma maculatum).*

Life History

The Spotted Salamander moves to the breeding ponds in winter to early spring. Migration occurs during rains over a period of a few days, and several migration events may ensue during the breeding season. Most egg deposition occurs from early January to mid-April during normal years, although unusually cold or warm weather can delay or accelerate reproduction. These salamanders have been observed in ponds in the Great Smokies as early as 10 January (in 1940); the earliest USGS field crews observed egg deposition (1998 to 2001) was 2 February. Huheey and Stupka (1967) reported that egg laying commenced nine years in January and seven years in February from 1937 to 1952. Egg masses have been seen as late as 7 May. Huheey and Stupka (1967) also noted up to 300 egg masses in the shallow pool at The Sinks on 17 March 1940; our counts have generally been less: 2 to 70 egg masses, depending on pool size.

Larvae are small (12–17 mm) when they hatch. The earliest date larvae have been seen is 22 March. Larvae of various size classes can appear bottomland on the Tennessee side of the park. Specific breeding sites include Gum Swamp, Gourley Pond, Abrams Creek Sinkhole Pond (near the Abrams Falls parking lot), and the Finley-Cane sinkhole ponds in Big Spring Cove.

Life History

Marbled Salamanders move to breeding ponds in the autumn and sequester themselves under debris in dry or moist pond basins. When fall and winter rains arrive, mating occurs and females deposit their egg clutches under surface debris. As the pond basin fills, the eggs develop and hatch directly into winter pools, where the larvae will grow and eventually metamorphose in early spring. Although other *Ambystoma* may use the ponds, the fall breeding of the Marbled Salamander ensures that their larvae will always be the largest salamander larvae in the breeding pools. This guarantees them a ready food supply as long as the pond fills. In times of fall and winter drought, however, Marbled Salamanders may have to skip breeding seasons.

Adults are present under logs and debris on the moist pond bottoms throughout the autumn and winter, temperature permitting. By mid-September to mid-October, the adults have completed their migration and are awaiting the rains. I have seen large gravid female Marbled Salamanders in the pond basin at Gum Swamp as early as 7 September 1998. We have not observed eggs in late September surveys. Huheey and Stupka (1967) reported eggs from 5 October to 16 December. Females deposit from 99 to 172 eggs per nest (King 1939b); several females may use the same cover object, but nesting is not communal. Depending of course when the rains arrive, eggs hatch into small larvae. USGS field crews observed a few larvae in Gum Swamp as early as 8 March 1999, but by mid-May the larvae were large and ready to metamorphose. Recent metamorphs have been found at Gum Swamp on 7 June, and both a few juveniles and adults may remain within the pond basin until early July.

Abundance and Status

Marbled Salamanders are rarely encountered except when they migrate to breeding ponds to await the rain, and while brooding eggs. Numbers undoubtedly vary annually, but there is no indication that these salamanders are experiencing stress within the park. They are vulnerable to disturbance while brooding, so breeding sites should be disturbed as little as possible during the autumn. If pigs entered the breeding sites at this time, the results could be disastrous. Some animals may be killed as they cross roads on their way to breeding ponds, particularly in the vicinity of Gum Swamp. However, most migration occurs at night, when this road is closed to vehicular traffic.

Remarks

As with other *Ambystoma,* this species has noxious skin secretions that help to deter small predators. It does not have a defensive tail display, but instead arches its body and "head butts" toward a would-be predator.

Mole Salamander Large juvenile, Gum Swamp (June 1998).

Mole Salamander
(Ambystoma talpoideum).

Mole Salamander
Ambystoma talpoideum

Etymology

Ambystoma: from the phrase *ana stoma buein,* to cram into the mouth; *talpoideum:* from the Latin *talpa,* meaning mole, and *oides,* meaning like. Literally, like a mole. The scientific name recalls a mole-like salamander with a large appetite.

Identification

ADULTS. Adults are stocky, and grayish to blue-gray to blackish in color. They have generally large heads, short tails, and large limbs. The dorsal surface is often flecked with whitish to blue specks. Juveniles are brown to greenish and retain the larval pattern for several weeks after metamorphosis, at which time they begin to develop the adult pattern. Adults attain sizes of 8 to 12 cm TL (3.15 to 4.75 in).

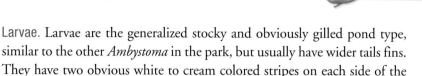

Larvae. Larvae are the generalized stocky and obviously gilled pond type, similar to the other *Ambystoma* in the park, but usually have wider tails fins. They have two obvious white to cream colored stripes on each side of the body that do not extend onto the tail; the lower of the stripes may be more distinct than the upper stripe. The abdominal vein is conspicuous on the belly, forming a dark midline stripe that persists in juveniles and adults.

Eggs. The exact method of egg deposition in the Great Smokies is unknown. On the Gulf Coastal Plain, eggs are deposited in small clusters on submerged pond debris, such as sticks and branches. Egg masses measure ca. 30 to 57 mm by 24 to 34 mm. Usually 3–4 up to 50 eggs are contained in each mass, and females deposit multiple clutches in a breeding season. On the Atlantic Coastal Plain, eggs are deposited singly, and a female may lay several hundred eggs per season. Older and larger females deposit more eggs than younger and smaller females.

Similar Species. The adults of this species are likely to be confused only with Marbled or Spotted Salamanders that lack their normal pigmentation. Marbled Salamanders have black bellies, whereas Spotted Salamanders are much larger and not as chunky as Mole Salamanders. Adult Mole Salamanders are much more bluish in background coloration than the former species. Marbled Salamanders transform in very early spring and have very dark to black bellies; the abdominal vein is not apparent. Mole Salamanders usually are lighter in color (blueish gray), have light to medium gray bellies, and have a dark line (abdominal vein) clearly visible running the length of the body. Spotted Salamanders transform much later than the former species, are medium to dark gray, usually lack the abdominal vein, and often have gold flecking dorsally that will eventually become the orange and yellow spots.

Taxonomic Comments. No subspecies are recognized.

Distribution

Mole salamanders are found from Virginia south to northern Florida, and westward to eastern Texas and Oklahoma. Populations occur in the delta lowlands of the Mississippi River north to southern Illinois. In the northern portions of its range, particularly in the mountains, the species occurs very discontinuously.

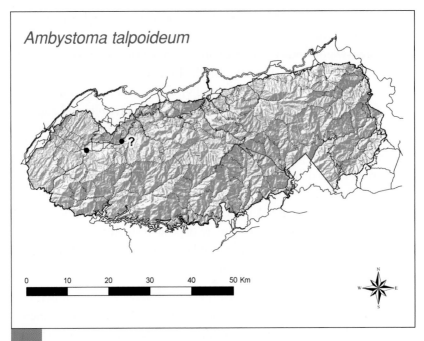

Distribution of the Mole Salamander *(Ambystoma talpoideum)*. The "?" marks a questionable locality in Big Spring Cove.

Mole salamanders are known from the Great Smokies only from the west side of Gum Swamp in Cades Cove, where two large subadults were found on 9 June 1998 (Dodd and Griffey 1999). Several larvae and a recently transformed juvenile, believed to be this species, were collected on 4 July 1998 at the Finley-Cane sinkhole ponds in Big Spring Cove (Dodd and Griffey 1999), but they were misidentified larval *A. maculatum*. Further attempts to collect mole salamanders at Gum Swamp have been unsuccessful; this species must be rare or very secretive within the park.

Life History

Nothing is known of the life history of this species within the park. Very large subadults were found at Gum Swamp in early July 1998, suggesting breeding in the early spring. At other locations, adults migrate to breeding ponds during cold rainy nights from late autumn to March. They emigrate from ponds in the early spring, and probably travel less than 200 m (333 ft)

Eastern Hellbender
Cryptobranchus alleganiensis

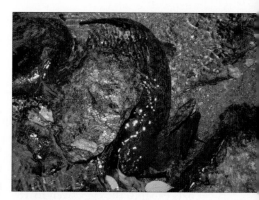
Hellbender adult, Little River.

Etymology

Cryptobranchus: from the Greek *kryptos,* meaning hidden, and *branchion,* meaning gills. The scientific name refers to the lack of large external gills; *alleganiensis:* Latinized proper name in reference to the Allegheny Mountains, where it was first discovered.

Identification

Adults. This is one of the largest salamanders in the world, and by weight it is the largest salamander in North America. The Hellbender has a somewhat flattened body, which allows access to crevices under large rocks and boulders, stout limbs, a flattened head with two small eyes, and a powerful tail. The most conspicuous of its characteristics are the fleshy folds that lie laterally on the side of the body and allow for cutaneous respiration. The body color is uniformly chocolate brown with mostly small dark spots or patches covering the entire body. Adults reach 30–74 cm TL (12–30 in). To the touch, Hellbenders feel particularly slimy, and, as such, they are difficult to handle.

Hellbender large juvenile, Little River.

Hellbender
(Cryptobranchus alleganiensis).

Larvae. Very small larvae are dark dorsally but have white bellies; six-month-old larvae start to develop the characteristic dark spots and blotches.

Hellbenders undergo only partial metamorphosis. As the external gills gradually are directed internally, with a single gill opening, the juveniles begin to resemble miniature adults.

Eggs. The female deposits between ca. 130 and sometimes more than 500 eggs in a nest under a large flat rock in the middle of a stream. The number of eggs varies geographically, but a somewhat low number of eggs in a nest, in comparison to the number that a female is capable of laying, may reflect cannibalism. The eggs are laid in two rosarylike strings that form a pile in the nest, and each measures 5–7 mm in diameter. After the jelly coats expand, the eggs swell to 18–20 mm in diameter, and are connected to one another by short cords.

Similar Species. No other species in North America resembles the Hellbender.

Taxonomic Comments: There are two subspecies of Hellbenders. *Cryptobranchus a. alleganiensis* occurs in the eastern United States, including the Great Smokies, whereas the Ozark Hellbender, *C. a. bishopi,* occurs in the Ozark highlands on the Missouri–Arkansas border.

Distribution

The Eastern Hellbender occurs from New York south to northern Georgia, Alabama, and northeast Mississippi, including most of the Alleghenies and Cumberland Plateau. A separate disjunct population is found in central Missouri.

Based on surveys from 1999 to 2000, Eastern Hellbenders are common in the Great Smokies of Tennessee only in the Little River, particularly from several miles east of its junction with Middle Prong of the Little River to where it exits the park near Townsend. In North Carolina, Eastern Hellbenders are known from the lower reaches of Deep Creek (between the park entrance and the campground) and in the Oconaluftee River (near the Floyd Cemetery). Unverified reports occur from Hesse Creek and Cosby Creek, both in Tennessee. Surveys in 2000 could not find Eastern Hellbenders in the Little Pigeon River, Cosby Creek, Big Creek, the Middle Prong of Little River, Abrams Creek, Fighting Creek, the West Prong of Little Pigeon River, Noland Creek, Hazel Creek, or Cataloochee Creek (Nickerson, Krysko, and Owen 2002).

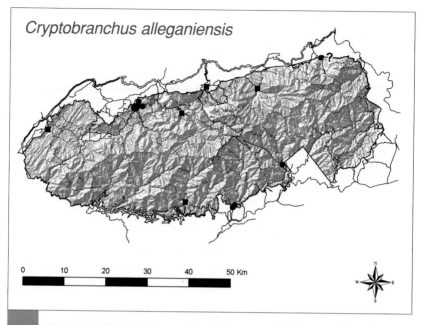

Distribution of the Hellbender *(Cryptobranchus alleganiensis)*. The squares show historic locations in Abrams Creek (left), Elkmont (top center-left), Gatlinburg (top center-center), Greenbrier (top center-right), Forney Creek (bottom center). The "?" indicates an unconfirmed report from Cosby Creek.

King (1939b) and Huheey and Stupka (1967) reported individuals from the Oconaluftee River near Smokemont (in 1944 and 1945), in the Little River near Elkmont (in 1936), and from the West Prong of the Little Pigeon River in Gatlinburg (in 1936, 1937, and 1942). Other records based on specimens in the National Park Service collection in the Great Smokies include Abrams Creek (captured by Wes Ogle on 7 August 1951), Forney Creek (no date, but a female with ovarian eggs), the Little Pigeon River at the former Greenbrier campground (July 1968), and in the Little Tennessee River (6 June 1959).

Life History

Little is known concerning the life history of this species within the Great Smokies. A single nest containing 140 near-term eggs was found in Little River on 15 October 2000; numerous larvae were seen at this general location from late August to mid-October (Nickerson, Krysko, and Owen

2002). This suggests that egg deposition in the Smokies occurs as early as late August. Males are territorial and will defend their cover sites. Breeding occurs in September and October, with egg deposition occurring shortly thereafter. After egg deposition, the female leaves the nest, and the eggs are guarded by the male. Incubation takes ca. 45 to 75 days, and hatching takes place in the autumn and winter. As noted above, egg eating is common in this species, including by the guarding male and the egg-laying female. The normal diet consists of small fish, aquatic insects, and especially crayfish. Hellbenders likely reach sexual maturity in 5 to 8 years, and may live 25 to 30 years in the wild.

Abundance and Status

Hellbenders are increasingly threatened throughout their range by a deterioration in water quality and by overcollecting. In the Smokies, the population in Little River seems reasonably healthy, with reproduction taking place and with variable size classes present. However, the section of the river inhabited by Hellbenders receives very heavy visitor use in summer. It is likely that Hellbenders are sometimes killed by swimmers and by fishermen, as they will occasionally take cast bait. The status of Hellbenders in other park streams is less certain, particularly in the Oconaluftee River and in Deep Creek. It is likely that the population in Abrams Creek was extirpated when the river was poisoned prior to stocking rainbow trout in 1957. Future management might include restocking Hellbenders into Abrams Creek. The population in the Little Tennessee River was probably extirpated during the filling of Fontana Lake.

Remarks

Hellbenders are often mischaracterized as "ugly" or "poisonous," and therefore killed on sight. It is not unusual to see newspaper reports picturing one of these salamanders as being "never before seen" or "unknown to science." While it is true that some people might find them ugly, that they will bite to defend themselves, that they are slimy, and that they contain noxious skin secretions, they are still magnificent ancient inhabitants of the streams of the Southern Appalachians. Hellbenders should be observed in their native habitat and left unmolested if found. They are not dangerous to people, will not bite unless provoked, and do no damage to native fish populations.

Seepage Salamander
Desmognathus aeneus

Etymology

Seepage Salamander adult,
Wolf Mountain Trail.

Desmognathus: from the Greek *desmos,* meaning ligament, and *gnathos,* meaning jaw. The name refers to the massive bundle of ligaments holding the jaw that gives a bulging appearance to the sides of the head of dusky salamanders; *aeneus:* from the Latin *aeneus,* meaning copper or bronze, presumably referring to the reddish dorsal red stripe.

Identification

Adults. The Seepage Salamander is a small, slender salamander with a normally straight-edged, dorsal, red stripe. The stripe may vary from yellow to brownish red, and it is bordered by a dark line. Sometimes a slight chevron pattern is present. The belly is mottled and light in color. Adults measure from 38 to 57 mm TL (1.5 to 2.25 in).

Larvae. None. This species experiences direct development, and the young normally hatch resembling miniature (10–12 mm [0.4–0.5 in] TL) adults. The gills that form during development in the egg are resorbed before hatching, although some embryos hatch with tiny remnant gills that are quickly resorbed.

Eggs. The eggs are white, surrounded by two jelly envelopes, and laid together in small clusters. They are small, befitting the size of the mother, and measure 2.4–3.0 mm in diameter including the surrounding capsule. Although the number of eggs deposited by females in the Great Smokies is unknown, they deposit between 5 and 17 eggs per clutch in other parts of their range. Communal nest site selection has been recorded, and, as with many salamander species, the female broods her eggs.

Similar Species. The Seepage Salamander somewhat resembles *Plethodon serratus* as well as some color patterns found in *Desmognathus conanti* and

D. ocoee. Southern Red-backed Salamanders are more terrestrial than Seepage Salamanders, and they do not possess the light line from the eye to the back of the jaw characteristic of salamanders of the genus *Desmognathus. Desmognathus ocoee* is more robust than the slender *D. aeneus,* and their ranges in the Smokies do not overlap; *D. ocoee* is found only at the highest elevations. Small *D. conanti* occasionally have a straight red stripe and are found in similar habitats as *D. aeneus;* separating them can be difficult. Seepage Salamanders usually have a Y-shaped mark on the head behind the eyes, and a light area on the dorsal surface of the thighs. Spotted Dusky Salamanders have a keeled tail in cross section, whereas Seepage Salamanders have a rounded tail. Pigmy Salamanders *(D. wrighti)* normally have a well-developed chevron pattern on their dorsum; unlike Seepage Salamanders, their heads are more rugose than smooth.

Taxonomic Comments: No subspecies are currently recognized, although at one time the populations in western Alabama were accorded separate subspecific status.

Distribution

The Seepage Salamander occurs from the Great Smoky Mountains south to east-central Alabama; there is also a disjunct population in western Alabama. In the Great Smokies, this small salamander occurs on the North Carolina side of the Park in the Hazel Creek (between Proctor and Walker Creeks, in Bone Valley), Eagle Creek (just north of Horseshoe Bend), and Twentymile Creek (along Wolf Ridge and Twentymile Loop Trails, Proctor Fields Gap) drainages. On 3 October 2000, Jeff Corser found a Seepage Salamander just inside the Great Smoky Mountains National Park boundary at U.S. 129 where the highway crosses into Tennessee; this is the first record for the Tennessee side of the park. On 14 May 2001, Corser found another locality along Pinnacle Creek (west of the Jenkins Ridge Trail at ca. 860 m [2,820 ft]), a small tributary of Eagle Creek. The species does not appear to occur farther east than Welch Ridge (i.e., into the Forney Creek or Noland Creek drainages).

Life History

In the Great Smokies, Seepage Salamanders are found under wet leaf litter, rocks, and surface debris along steep road banks and trail sides; true to its name, it is usually found in seepage areas that rarely if ever dry completely.

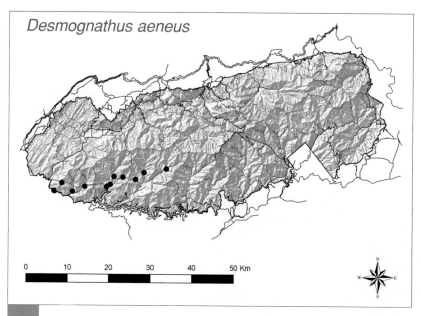

Distribution of the Seepage Salamander *(Desmognathus aeneus).*

Nothing is known of the life history of this species within the park. All specimens observed have been adults, with the exception of a very small animal found by Jeff Corser on 8 October 1999 along the Twentymile Loop Trail. Mating probably occurs in both the autumn and spring, and eggs are deposited in nests in mosses or wet, rotten logs near seepages. Eggs are laid in April to May, and hatching occurs from late May to August. These tiny salamanders eat a wide variety of invertebrates, particularly mites, isopods, and amphipods.

Abundance and Status

Seepage Salamanders seem reasonably common within their restricted habitats in the park. However, roadbank modification, such as might occur along the Hazel Creek road, could have adverse impacts on seepage habitats.

Remarks

Seepage Salamanders were first collected in the Great Smokies along Hazel Creek by Richard Highton of the University of Maryland in the early 1960s, although the location was never published. Surveys in 1998 reconfirmed the presence of Seepage Salamanders at Highton's original collecting sites.

When uncovered in the field, these salamanders often freeze in position, termed "immobility," presumably making them difficult to see by a visually oriented predator attracted by movement (Dodd 1990). Their red stripe and small size make them blend into the background and thus surprisingly difficult to see among the reddish colored leaf litter and small rocks in seepage habitats.

Spotted Dusky Salamander adult, Fighting Creek.

Spotted Dusky Salamander
Desmognathus conanti

Etymology

Desmognathus: from the Greek *desmos,* meaning ligament, and *gnathos,* meaning jaw. The name refers to the massive bundle of ligaments holding the jaw that gives a bulging appearance to the sides of the head of dusky salamanders; *conanti:* Latinized version of Conant, in honor of famed herpetologist Roger Conant.

Identification

Adults. The Spotted Dusky Salamander is a medium-sized desmognathine salamander (see *D. imitator* account) with an extremely variable dorsal pattern. The background color is tan to brown to nearly black, and older, large individuals may be completely melanistic. The pattern may include a wavy stripe of brown to yellow to red, or it may be blotched or nearly uniform. The sides are dark colored, and the side and bellies are salt-and-peppered with white flecks. Bellies are light to dark gray, but not black, and are often salt-and-peppered. The tails are moderately keeled. In juveniles, the spots fuse to form a stripe or blotched adult pattern. Adults measure 6–11 cm (2.4–4.5 in) TL.

Larvae. The larvae are small and brown, and have from 5 to 8 alternating pairs of spots between the limbs when viewed from above. Spots continue onto the tail.

Eggs. Eggs are cream-colored to white, and measure 2.5 to 3 mm in diameter. Clutch size has not been determined in the Great Smoky Mountains, but other Fuscus complex desmognathines of similar size deposit ca. 20 eggs per clutch.

Similar Species. When attempting to identify a medium-sized dusky salamander, read the various species descriptions and look at the illustrations for comparison. Santeetlah Salamanders usually have a light yellow wash on the underside of the

Spotted Dusky Salamander
(Desmognathus conanti).

limbs and tail (Spotted Duskies do not). Imitators also have dark or black bellies. The pattern of Spotted Duskies is much brighter and bolder than that of the Santeetlah Salamander. According to Tilley (1981), Santeetlah Salamanders usually remain motionless when disturbed, making them easy to catch, whereas Spotted Duskies move vigorously through muck and water to escape capture. The larvae of some small desmognathines may be virtually impossible to distinguish from one another. One helpful character is the number of paired or alternating spots between the limbs (see table 3).

Taxonomic Comments. The nomenclature surrounding the salamanders of the Fuscus complex of the genus *Desmognathus* in the Great Smoky Mountains is problematic to say the least. The Spotted Dusky is treated as a species by Titus and Larson (1996) and as a subspecies of *D. fuscus* by Petranka (1998). Even if it is considered a subspecies of *D. fuscus,* there is some confusion as to whether Smokies populations should be referred to *D. f. fuscus* or *D. f. conanti.* It appears that there is a broad contact zone between *D. fuscus* and *D. conanti* in the vicinity of the Smokies. Petranka (1998) noted that genetic variation is poorly documented in the region. *Desmognathus conanti* hybridizes extensively with another member of the Fuscus complex, *D. santeetlah,* throughout the northwest side of the park in Tennessee, further confusing identification. The great degree of individual color and pattern variation also is of little help in identification.

Recognizing the confusion concerning nomenclature, I have chosen to follow Titus and Larson (1996) in referring the all mid- to low-elevation dusky salamanders within Great Smoky Mountains National Park, exclusive of *D. santeetlah,* to *D. conanti.* Clearly, more research, using molecular techniques, must be conducted to sort out the species identification(s) of the dusky salamanders within the Great Smokies.

Distribution

The Spotted Dusky Salamander occurs in a broad band from western Kentucky through Tennessee to the region of the Great Smoky Mountains, and eastward to the Fall Line in South Carolina. The species occurs from the broad area outlined above southwestward to the Gulf Coast, with isolated populations in north-central Louisiana and adjacent Arkansas. The Spotted Dusky Salamander, or at least members of this portion of the Fuscus complex, occurs at lower elevations throughout the Park. Most Spotted Dusky Salamanders were collected below an elevation of ca. 792 m (2,600 ft). The highest elevation it was collected during surveys from 1998 to 2001 was 960 m (3,150 ft), just southwest of the Little Cataloochee Church.

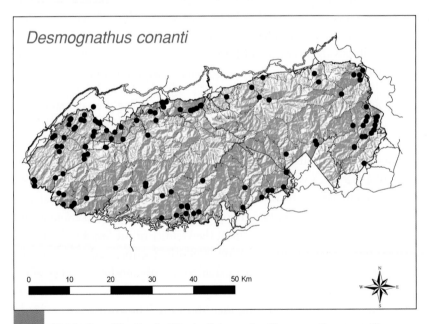

Desmognathus conanti

Distribution of the Spotted Dusky Salamander *(Desmognathus conanti).*

Life History

There are no data specifically from the Great Smoky Mountains on the life history of this species. The Spotted Dusky Salamander occurs along small lowland streams and in seepage areas. Activity occurs both by day and night. Nests probably are placed under moss and litter in seepages and streams, and are guarded by an attending female. Eggs are deposited in the summer, with larvae hatching from summer into autumn. The larval period probably lasts about one year.

Abundance and Status

The Spotted Dusky Salamander appears to be extremely common at lower elevations. Nothing is known about how human activities affect this species, especially the impacts of atmospheric pollution.

Remarks

If ever the systematic status and biology of a salamander deserved clarification in the Great Smoky Mountains, the Spotted Dusky Salamander is that species.

Imitator Salamander
Desmognathus imitator

Etymology

Desmognathus: from the Greek *desmos,* meaning ligament, and *gnathos,* meaning jaw. The name refers to the massive bundle of ligaments holding the jaw that gives a bulging appearance to the sides of the head of dusky salamanders; *imitator:* refers to the mimicry by many individuals of this species to sympatric red-cheeked *Plethodon jordani* (see below).

Identification

Adults. The Imitator Salamander is a medium-sized desmognathine salamander, meaning that it can be identified, in part, to genus by observing a light line running from the eye to the posterior angle of the jaw. Its dorsum is brown to nearly black, and most Smokies animals have dark to nearly black bellies. A faint, dark pattern sometimes is visible which may appear

Imitator Salamander adult, with red cheeks. Beech Flats at Luftee Gap.

Imitator Salamander adult, dark phase without red cheeks. Indian Gap.

Imitator Salamander adult, unusual red coloration. Heintooga Road.

as a wavy, but not straight, dark brown stripe. The tail is rounded in cross-section. Any *Desmognathus* in the Great Smokies with yellow to red cheeks belongs to this species. Adults measure 7–11 cm TL (2.75–4.3 in).

Larvae. Larvae have a dark background color with distinct dorsal orange to chestnut-colored spots. They are small, ca. 10 to 12 mm SVL, and may have clearly visible cheek patches.

Eggs. The eggs are cream-colored and deposited in mono- or bilayer clusters underneath debris and moss in seepage areas. Clutch size ranges from 13 to 30, and eggs measure 3.5 mm in diameter (Koenings et al. 2000). Females brood their eggs.

Similar Species. In the areas of contact or sympatry, uniformly dark-colored Imitator Salamanders may be easily confused with older Spotted Dusky Salamanders *(D. conanti)*, Santeetlah Salamanders *(D. santeetlah)*, and Ocoee Salamanders *(D. ocoee)*. In some cases, even experts may be unable to identify a specimen without the aide of molecular data. When attempting to identify a medium-sized dusky salamander, read the various species descriptions and look at the illustrations for comparison. Note the location and elevation where the individual was found. Ocoee Salamanders usually have a straight-edge dorsal stripe (Imitators do not); Santeetlah Salamanders usually have a light yellow wash on the underside of the limbs and tail (Imitators do not); Spotted Dusky Salamanders usually have grayish salt-and-pepper bellies, and their tails are keeled in cross-section (Imitators have black bellies and rounded tails in cross-section). Unfortunately, older animals, in particular, among all four species tend to become uniformly dark.

Taxonomic Comments. *Desmognathus aureatagulus,* described by Weller (1930a) based on a yellow-cheeked *Desmognathus* from near Newfound Gap, is actually this species. The Waterrock Knob form of the Imitator Salamander differs phenotypically, ecologically, and genetically from "typical" *D. imitator* in the Great Smokies and Balsams. Tilley (2000) suggested that there were valid reasons for recognizing this form as a different species, but declined to do so "because it would obscure major patterns of evolutionary diversification." In general, *D. imitator* displays substantial genetic differences among its populations in the Smokies and Balsams despite the overall restricted geographic range of this species.

Imitator Salamander
(Desmognathus imitator).

Distribution

The Imitator Salamander is found only in the Great Smoky Mountains, the Balsam Mountains, and on nearby Waterrock Knob (Plott Balsam Mountains) along the Blue Ridge Parkway (Tilley 2000). This is generally a high-elevation species (greater than 900 m, according to Petranka 1998), occurring to the top of Clingmans Dome. Most records during field surveys from 1998 to 2001 were at elevations greater than 1,280 m (4,200 ft). However, Imitator Salamanders were found at several lower-elevation sites, including the north slope of Maddron Bald (756 m [2,480 ft], 866 m [2,840 ft]) in Tennessee, and in the Solola Valley (776 m [2,545 ft]), along Noland Creek near Jim Ute Branch (841 m [2,760 ft]), and on the west slope of Richland Mountain above Kephart Prong (1,110 m [3,640 ft]) in North Carolina. The isolated Waterrock Knob form occurs only at elevations between 1,650 m and 1,800 m (5,410 ft and 5,905 ft).

Life History

Imitator Salamanders are found along streams and in seeps, but they are also found far out onto the forest floor and well away from water. They live

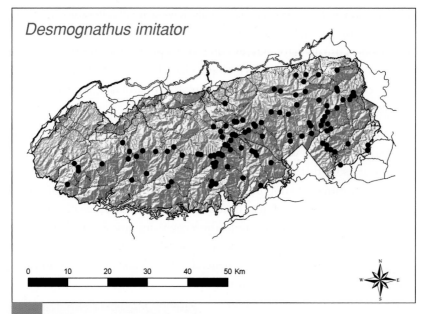

Distribution of the Imitator Salamander *(Desmognathus imitator)*.

under surface debris and amongst leaf litter. There is some information in the literature on reproduction of dusky salamanders within the park (King 1939b; Wood and Wood 1955), but the older literature must be used with caution because of considerable confusion in the identification of medium-sized *Desmognathus* in the Smokies. The Imitator Salamander deposits eggs under moss or in the interstices between rocks in small rivulets and head-water seepage areas (Bernardo 2000; Koenings et al. 2000). I have observed gravid females on 3–8 June near the Heintooga Overlook, on Clingmans Dome, and along Beech Flats Creek. Five clutches of eggs ($\bar{x}$ = 19 eggs per clutch) were found on 30 June 1998 near Heintooga Overlook (Koenings et al. 2000). However, gravid females also have been captured as late as 8 July (1999), 25 July (2000), and 2–3 August (1999, 2000). Embryos hatch as tiny larvae, but little is known about the early life of this species. Larvae likely transform after 9 to 10 months, as do similarly sized sympatric *D. ocoee*. A singular tiny animal in the process of metamorphosis was seen on 7 July (2000) on Clingmans Dome, and Bernardo (2000) reported 54 tiny recent metamorphs at a small seepage on the trail to Mount Sterling on 13 July (1993). Small juveniles are observed from July through September.

The food of this salamander consists of any small invertebrate that it can catch. In the Smokies, I have watched adult Imitator Salamander capture both worms and centipedes. They are usually considered to be sit-and-wait predators, although I have observed them commonly out walking on the surface litter in the middle of the day; whether they were hunting or simply moving between cover sites is unknown. Imitator Salamanders occasionally even climb trees during the day; I once encountered an individual 75 cm high on a tree near the Heintooga Overlook, and another under bark 2 m up a tree along Roaring Fork.

Abundance and Status

The Imitator Salamander is abundant in all habitats where it is found. In the Great Smokies, Imitator Salamanders are found in significantly higher numbers on sites without a previous history of disturbance by farming or logging (Hyde and Simons 2001). At present, there are no indications that this endemic species is threatened within the Great Smoky Mountains.

Remarks

One of the best examples of Batesian mimicry in the animal world is found between this species, which is palatable to predators, and Jordan's Salamander, which is highly distasteful and avoided by most predators (Brodie and Howard 1973). Imitator Salamanders often have bright red to yellowish cheeks (15–20 percent of the population; Tilley et al. 1978), which are very similar to the bright red cheeks of Jordan's Salamander. Predators, particularly birds, can learn to avoid the distasteful, noxious Jordan's Salamander, and will likewise avoid the palatable but similarly colored Imitator Salamander (birds, especially, are visually oriented predators). As might be expected, red-cheeked Imitator Salamanders are usually less abundant than their red-cheeked models. By being less abundant (and therefore encountered less frequently) than Jordan's Salamanders, visually oriented predators are unlikely to learn to recognize the mimics. Mimicry between a palatable *Desmognathus* and a distasteful *Plethodon* occurs in many areas of the southern Appalachians. Although red-cheeked *P. jordani* only occur in the Great Smokies, red-cheeked *Desmognathus* of other species occasionally are found elsewhere. The reason for this is unknown, although the possession of a bright color might be a widely recognized warning signal to predators that the owner *may* be inedible. Or maybe not: on three occasions between 1998 and 2000, I found freshly dead specimens in the field, most likely indicating a failed predation attempt. Imitator Salamanders easily lose their tails, and animals have been seen in the Smokies with bite marks or missing limbs.

Shovel-nosed Salamander adult, Mill Creek in Cades Cove.

Shovel-nosed Salamander larva, Balsam Mountains near Whim Knob.

Shovel-nosed Salamander
(*Desmognathus marmoratus*).

Shovel-nosed Salamander
Desmognathus marmoratus

Etymology

Desmognathus: from the Greek *desmos,* meaning ligament, and *gnathos,* meaning jaw. The name refers to the massive bundle of ligaments holding the jaw that gives a bulging appearance to the sides of the head of dusky salamanders; *marmoratus:* from the Latin *marmoratus,* meaning marbled, presumably in reference to its often marbled dorsal pattern.

Identification

Adults. This nearly entirely aquatic species is light to very dark brown dorsally. Two rows of small white spots normally are found laterally on the body. Some individuals in the Smokies have yellowish brown splotches on the body and on the dorsal crest of the tail fin. The yellowish brown coloration also is observed in younger animals. Heads may be lighter in color than the body. The bellies are light in young animals, but become dark gray to even black in older animals, at least in the Great Smokies. Tails are laterally compressed, and the toe tips have dark cornified pads to assist walking in swift water. The heads are shovel-shaped and more dorso-ventrally compressed than the heads of Black-bellied Salamanders. Adults measure 8–15 cm TL (3.15–6 in).

Larvae.. The larvae are dark colored with conspicuous white flecks along the side of the body. They are slimmer than the larvae of Black-bellied Salamanders, which they resemble, and they have relatively longer legs and a spatulate (as opposed to a pointed) tail. However, tail shape is often not a very reliable character, due perhaps to abrasion in swift waters or predation attempts. The gills are white, as they are in other *Desmognathus*. Conspicuous dorsal spots of red or orange are not present.

Eggs. Unpigmented eggs are attached to the underside of rocks in swift-flowing water. They are attached singly or in clusters of 2 to 4. Females deposit 20–65 eggs, with larger females producing larger clutches. The eggs are 4–5 mm in diameter and surrounded by a jelly capsule. The developmental period lasts 10–12 weeks.

Similar Species. This species very closely resembles the Black-bellied Salamander, which often lives in the same streams as the Shovel-nosed Salamander. Black-bellied Salamanders usually are more robust, have completely black bellies, possess "buggier" eyes, and have internal nares (openings in the roof of the mouth) that are round; in contrast, Shovel-nosed Salamanders are somewhat slimmer, may have gray or lighter colored bellies, have less prominent eyes and a more flattened head, and the internal nares are slits. Another difference involves behavior. When uncovered, Shovel-nosed Salamanders often sit still or very quietly walk away on the stream bottom. Black-bellied Salamanders are quick as lightning in their escape attempts, and generally do not remain calm when something is pursuing them.

Taxonomic Comments. This species was formerly placed in its own genus, *Leurognathus,* but molecular evidence suggests that it is closely related to *D. quadramaculatus* and does not warrant separate generic status (Titus and Larson 1996). Five different subspecies, based on size, color, and other minor morphological differences, have been named, including *L. m. melania* from the Great Smokies (Martof 1956). Until more data are available, it seems prudent not to recognize any subspecies, although populations from different drainages yet may prove to be different from one another at the molecular level.

Distribution

The Shovel-nosed Salamander occurs in cold mountain streams of the Southern Appalachians from southwestern Virginia to northeast Georgia. Isolated populations are found in southwest-central Virginia and in northern Georgia. The determination of the range of this species in the Great Smoky Mountains has been difficult because of ambiguities in discriminating Black-bellied Salamanders *(Desmognathus quadramaculatus)* from this species (see the Remarks heading below). Most localities recorded during field surveys from 1998 to 2001 were in lower-elevation streams, such as the Little Pigeon River, Cosby Creek, Porter's Creek, Mill Creek in Cades Cove, Kirkland Branch (flowing into Fontana Lake), Cold Spring Branch (a small tributary to Hazel Creek) and Rough Fork Creek in Cataloochee. Shovel-nosed Salamanders occur widely throughout the park, but appear to be scarce at higher elevations. There are records for ca. 1,500 m (4,920 ft)

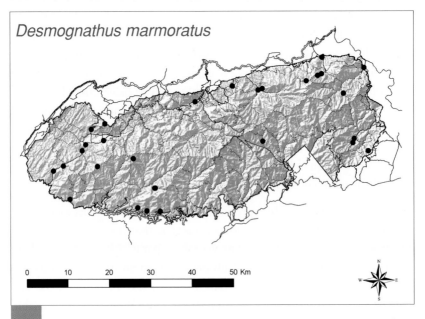

Desmognathus marmoratus

Distribution of the Shovel-nosed Salamander *(Desmognathus marmoratus)*.

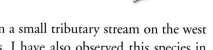

in Beech Creek at Luftee Gap and in a small tributary stream on the west side of Whim Knob in the Balsams. I have also observed this species in Kephart Prong at 1,000 m (3,280 ft). However, most locations are at elevations greater than 762 m (2,500 ft).

Life History

Nothing is known about the life history of this species in the Great Smokies. In other areas, females deposit their eggs in spring to early summer, and remain with them until hatching. Hatching occurs in August and September. Larval Shovel-nosed Salamanders are more active at night than during the day, and they take about three years to metamorphose into adults. Juveniles and adults co-occur in streams, where they hide under rocks and debris and forage along the stream bottom, usually at night. This species eats many different types of aquatic invertebrates, particularly mayfly and caddisfly larvae. As with many species of salamanders, they eat just about anything they can catch and swallow.

Abundance and Status

This species appears to be widespread and not uncommon within the Great Smoky Mountains National Park, although it is not nearly as numerous as the Black-bellied Salamander. The clear streams and fast-flowing cold waters provide ideal habitats. In other regions of the Southern Appalachians, siltation and pollution may have adversely affected Shovel-nosed Salamander populations.

Remarks

This species has been widely confused with the Black-bellied Salamander, leading perhaps to inaccurate information in the scientific literature concerning the population biology of large stream desmognathines in the Great Smoky Mountains. For example, Mathews (1984) considered the Shovel-nosed Salamander to be the most abundant stream salamander throughout the park, yet no other researcher working in the Smokies has found similar results. Data collected from 1998 to 2001 suggest that while this species is widespread, it is not abundant at most locations. Likewise, it is likely that the disappearance of larval Shovel-nosed Salamanders reported by Mathews and Morgan (1982) in connection with road construction through the Anakeesta Formation near Newfound Gap actually refers to the Black-bellied Salamander (see Kuchen et al. 1994). Biologists conducting surveys in the headwaters of Beech Flats Creek above the area affected have found only Black-bellied Salamanders (Kuchen et al. 1994; J. Petranka pers. comm.; USGS survey results 1998–2001, unpubl. data).

Seal Salamander adult, Greenbrier Cove.

Seal Salamander
(*Desmognathus monticola*).

Seal Salamander
Desmognathus monticola

Etymology

Desmognathus: from the Greek *desmos,* meaning ligament, and *gnathos,* meaning jaw. The name refers to the massive bundle of ligaments holding the jaw that gives a bulging appearance to the sides of the head of dusky salamanders; *monticola:* from the Latin *monticola,* meaning mountaineer or highlander, in reference to the species's mountain habitat.

Identification

Adults. This is a large tan to brown, somewhat robust salamander. There is often a series of black or dark brown markings on the dorsum, which can be seen as reticulations or irregular splotches. Patterns are quite variable. The bellies are light colored and unmottled, with a distinct demarcation in coloration between the sides and the belly. Tails are keeled toward the end, but rounded at the base. The toes are cornified at their tips. Adults measure 7.5–15 cm TL (3–6 in).

Larvae. The larvae are light brown with four or five pairs of dorsally located large reddish spots between the fore and hind limbs. As the larva grows, the spots fuse and blend into the black dorsal reticulations of the adult.

Eggs. Eggs are deposited under surface debris or rocks in or immediately adjacent to running water. They may be placed in wet, undercut stream banks. Clutch size varies geographically, with 16 to 40 eggs in the Southern Appalachians. Eggs are deposited singly, attached to cover via an elastic stalk, and appear in layers of twos and threes within the nest. The egg size is ca. 4 mm in diameter. Females sometimes guard their clutches.

Similar Species. There are three species of large *Desmognathus* in the Smokies that are somewhat similar in appearance. In contrast to Seal Salamanders, Black-bellied Salamanders are darker and have black bellies, and the heads of Shovel-nosed Salamanders are shaped differently and they have gray to dark bellies. Unlike larval Seal Salamanders, the larvae of these species do not have conspicuous reddish spots.

Taxonomic Comments. There is a great deal of pattern and color variation within this species, and at one time *D. monticola* was partitioned into two subspecies. However, only the single wide-ranging species is currently recognized. Once a molecular analysis has been carried out, taxonomic changes eventually may be warranted.

Distribution

The Seal Salamander occurs in mountainous regions from western Pennsylvania to east-central Alabama. There are disjunct populations in the Red Hills of southern Alabama and in some of the deep ravines of the Florida panhandle. The Seal Salamander occurs at generally mid to lower elevations throughout the Great Smoky Mountains. At higher elevations, the distribution becomes much more spotty. In field surveys from 1998 to 2001, *D. monticola* was found from 381 m (1,250 ft) along Cane Creek to 1,573 m (5,160 ft) at Luftee Gap and 1,646 m (5,400 ft) near the Heintooga Overlook picnic area. Most locations where this species was found were at elevations less than 762 m (2,500 ft), however. Huheey and Stupka (1967) recorded it at 1,676 m (5,500 ft).

Life History

Seal Salamanders occur most often in streams and along streambanks at lower elevations, and are frequently found associated with wet rock faces (e.g., at locations along Little River Road). Eggs are deposited from June

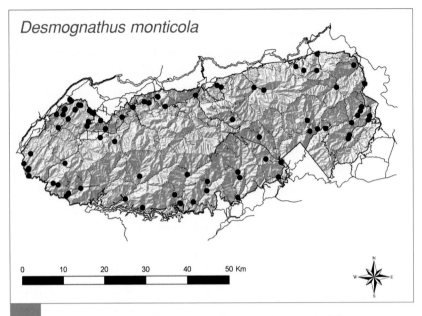

Distribution of the Seal Salamander *(Desmognathus monticola).*

to August, with hatching occurring in late summer to early autumn. After overwintering, they transform midway through the next summer, that is, from 8 to 11 months after hatching. Maturity occurs for males at 3.5 years, and for females at 4.5 years. Seal Salamanders may live to 11 years of age. As with most salamanders, Seal Salamanders eat about any small prey they can catch, from small invertebrates to other salamanders. They may even take up novel positions in order to catch prey: on 11 July 1999, I observed a large adult near Gourley Pond sitting on a fallen mushroom eating flies drawn to the rotten odor of the fungus.

Abundance and Status

Seal Salamanders are commonly found in various areas throughout the park, but they become scarcer at higher elevations. As long as water quality is maintained, they should do well within the Great Smoky Mountains National Park. However, as with most other stream-dwelling salamanders, little is known about how atmospheric pollution (both pollutants and acid rain) will affect this species in the future. Many aquatic salamander larvae cannot survive high levels of acidity.

Remarks

Although generally considered a wet seepage or streamside salamander, I have observed this species 3 m (10 ft) from water. In the lower elevations of the Smokies, they are the large *Desmognathus* commonly seen on wet rock faces.

Ocoee Salamander
Desmognathus ocoee

Etymology

Desmognathus: from the Greek *desmos,* meaning ligament, and *gnathos,* meaning jaw. The name refers to the massive bundle of ligaments holding the jaw that gives a bulging appearance to the sides of the head of dusky salamanders; *ocoee:* named for the Ocoee River Gorge in Tennessee where the species was first found.

Ocoee Salamander adult, Indian Gap. Note that the stripe may range from brown to gray, tan, yellow, or red.

Identification

Adults. The Ocoee Salamander is a medium-sized desmognathine salamander (see *D. imitator* account) that inhabits the highest elevations in the Great Smoky Mountains. In the Great Smokies, Ocoee Salamanders have a straight-edged dorsal stripe running the length of the body. The stripe varies from tan to brown, gray-green, yellow or red, and in older animals it may be nearly as black as the bordering line and background body color. In young animals, the belly is white, but becomes more

Ocoee Salamander recently transformed metamorph, Heintooga Overlook (July 8, 2000).

Ocoee Salamander *(Desmognathus ocoee).*

pigmented with age. However, it is usually not jet black as is the belly of *D. imitator.* Older animals may become very melanistic. Adults measure 7–11 cm TL (2.75–4.3 in).

Larvae. The larvae are very small and dark; in the Smokies, at least, some larvae have the striping pattern of the adult. Other larvae have 4–6 alternating light spots on the dorsal surface that break down and fuse to form the stripe. Larvae have rounded snouts.

Eggs. Eggs are deposited in small clusters under litter and moss in springs, seeps, and small streams. The female deposits between 5 and 23 eggs, each about 2 to 3 mm in diameter. Clutch size increases with female size, and larger clutches are found at higher elevations. Females brood their egg clutches, which prevents predation by beetles and other salamanders, inhibits fungal infection, and prevents contamination by infertile decaying eggs.

Similar Species. This species may be confused with *D. imitator* (most often) and *D. santeetlah.* It occurs at much higher elevations than *D. conanti.* See the discussion under the heading similar species in the Imitator Salamander account.

Taxonomic Comments. The Ocoee Salamander is a member of the Ochrophaeus complex of salamanders of the genus *Desmognathus,* an extremely morphologically and genetically variable group of salamanders that inhabits the Southern Appalachian Mountains. Without knowing exact location data (place, elevation, habitat), it may be impossible to put a name on a particular individual and, even then, molecular data may be required. Because of the morphological and pattern variation, even salamander experts have difficulty identifying individuals in the field. The taxonomic status of the Ocoee Salamander is reviewed by Tilley and Mahoney (1996) and Petranka (1998).

Distribution

The Ocoee Salamander occurs in western North Carolina and adjacent Tennessee, northwestern South Carolina, and northern Georgia. A dis-

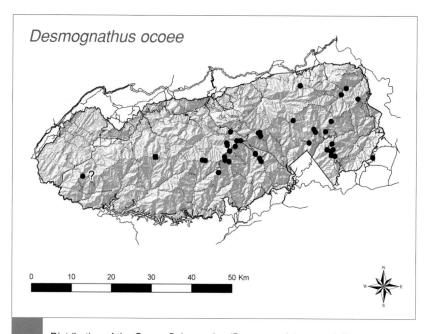

Desmognathus ocoee

| 0 | 10 | 20 | 30 | 40 | 50 Km |

Distribution of the Ocoee Salamander *(Desmognathus ocoee)*. The square shows the most western confirmed locality for this species, at Starkey Gap (see Tilley and Huheey 2001). The "?" indicates a location on Gregory's Bald where a salamander that resembled *D. ocoee* was found; no genetic analysis was carried out to confirm its identity, however.

junct population occurs in northeastern Alabama. In the Great Smoky and Balsam Mountains, Ocoee Salamanders are found only at the highest elevations of the park (at least from Starkey Gap in the west to Mount Sterling in the east; suspect record from Gregory's Bald in the west), although in other parts of their range they are found at much lower elevations. During field surveys from 1998 to 2001, this species was nearly always found at elevations from 1,463 to 1,830 m (4,800 to 6,000 ft). Salamanders matching the description of *D. ocoee* are occasionally found at lower elevations (e.g., at Hen Wallow Falls [866 m, 2,840 ft] and Albright Grove [1,012 m, 3,320 ft]). Inasmuch as certain salamanders of the lower elevation *D. fuscus* complex sometimes have a morphological pattern similar to Smokies *D. ocoee,* identification at middle and lower elevations can only be verified by using molecular techniques.

Life History

As with *D. imitator,* Ocoee Salamanders are found in seeps and trickles at higher elevations within the Great Smokies, and they also are quite terrestrial. I have observed them far from the nearest water under surface debris and leaf litter. Very little is known concerning the early life history of this species. Mating likely occurs from spring to autumn. Gravid females have been observed as early as 1 July (1998) near the Heintooga Overlook and as late as 19 September on Clingmans Dome. Transforming *D. ocoee* (at 10 mm SVL) were observed on 4 August (2000) near the Heintooga Overlook. In some populations in western North Carolina, hatching occurs in August and September, and the larval period lasts 9–10 months. Males and females reach maturity at ca. 28 mm SVL (1.1 in) at an age of 3 years. Longevity may reach 7–10 years. Insects and other small invertebrates form the prey of this species, and occasionally Ocoee Salamanders are the prey of larger salamanders. On 20 July 1999, Jeff Corser (pers. comm.) saw a large Black-bellied Salamander consume an Ocoee Salamander along the Maddron Bald Trail.

Abundance and Status

The Ocoee Salamander seems fairly common at higher elevations. However, nothing is known concerning its habitat requirements, particularly about how its biology is influenced by habitat structure. Exactly how changes to the forest (resulting from acid precipitation, pollution, or changes in vegetative species composition because of the woolly adelgid infestation) might affect this species is unknown.

Remarks

In its range outside the Great Smoky Mountains, the color pattern and favored habitats and elevations of this species are quite different. Other populations lack the straight stripe, are much more mottled, and possess wavy or blotched patterns. These differences are discussed in detail in Petranka (1998:197), who noted: "This species ranges over a greater elevational span than any other *Desmognathus* species and extends from low-lying gorges to the highest mountaintops in the Great Smoky Mountains."

Black-bellied Salamander
Desmognathus quadramaculatus

Etymology

Black-bellied Salamander adult, showing light-headed form. Parsons Branch.

Desmognathus: from the Greek *desmos,* meaning ligament, and *gnathos,* meaning jaw. The name refers to the massive bundle of ligaments holding the jaw that gives a bulging appearance to the sides of the head of dusky salamanders; *quadramaculatus:* from the Latin *quadratus,* meaning four-sided, and *maculatus,* meaning spotted.

Identification

Adults. This is a large, stout, and robust stream-dwelling salamander. The heads are large, due to the massive jaw muscles located on the sides of the head. The dorsal color is mottled dark brown to black, and some individuals have a weakly developed, rust-colored splotchy pattern. One or two rows of small white spots are located laterally.

Black-bellied Salamander large juvenile, "stardust" pattern. Beech Flats at Luftee Gap.

The bellies are jet black. Some adults, particularly around Newfound Gap, have a "stardust" pattern, with many light yellowish to gold flecks about the head and body. Juveniles resemble the adults, except that the pattern may be more noticeable. Younger juveniles have 6–8 pairs of spots between the limbs, dorsally; these eventually break down to form the splotched pattern. Juveniles are also more brown than are the old adults, and their bellies may be light. As the animal grows, the color changes to black. According to Petranka (1998), the bellies darken within a year of metamorphosis. Adults measure 9–21 cm (3.5–8.25 in) TL.

Black-bellied Salamander recent hatchling, showing large amount of yolk. Fighting Creek (August 2000).

Black-bellied Salamander (*Desmognathus quadramaculatus*).

Larvae. The larvae are brown to very dark colored, stocky, and large. The gills are white. They have cornified toe tips, the juvenile spotting pattern, and light bellies. These are the largest *Desmognathus* larvae in the Great Smokies.

Eggs. The female normally deposits 21–54 eggs under rocks in fast-flowing water, and a female will guard her egg clutch. However, a specimen collected in 1950 in the Smokies had 60 eggs (Huheey and Stupka 1967). Petranka (1998) found eggs of this species deposited in globular clusters, rather than in layers, in the Great Smokies. The eggs were in smaller groups, but attached to a main stalk via short side stalks, much like a cluster of grapes.

Similar Species. This species very closely resembles the Shovel-nosed Salamander, which often lives in the same streams as the Black-bellied Salamander. Black-bellied Salamanders usually are more robust, have completely black bellies, possess "buggier" eyes, and have internal nares (openings in the roof of the mouth) that are round. In contrast, Shovel-nosed Salamanders are somewhat slimmer, may have gray or lighter colored bellies, have less prominent eyes and a more flattened head, and the internal nares are slits. Another difference involves behavior. When uncovered, Shovel-nosed Salamanders often sit still or very quietly walk away on the stream bottom. Black-bellied Salamanders are quick as lightning in their escape attempts, and generally do not remain calm when something is pursuing them.

Taxonomic Comments. There is a degree of morphological variation in Black-bellied Salamanders from different parts of its range, and at one

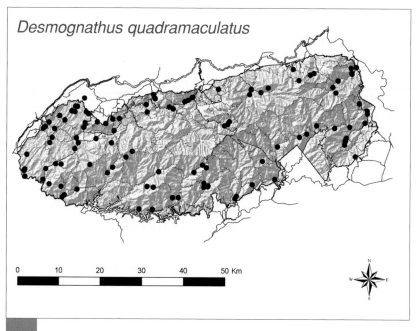

Distribution of the Black-bellied Salamander *(Desmognathus quadramaculatus)*.

time different subspecies were recognized. New molecular evidence confirms that Black-bellied Salamanders may not represent a single evolutionary lineage within the Southern Appalachians and Piedmont and that the genetic variation in some populations might be such as to elevate them to distinct species. Until such evidence is published, however, the "Black-bellied Salamander" is considered as a single species.

Distribution

The Black-bellied Salamander occurs from southern West Virginia to northern Georgia and South Carolina. There are several disjunct populations in west-central Georgia. In the Great Smoky Mountains, Black-bellies are found throughout the park. They occur at elevations from 341 m (1,120 ft) near the Abrams Creek campground to 1,714 m (5,625 ft) on Andrews Bald. Huheey and Stupka (1967) give an upper elevation of 1,829 m (6,000 ft).

Life History

This is a salamander of nearly all small to large streams in the Great Smokies, although it also may be found in seepages, in crevices near streams (e.g., along Le Conte Creek), on rock faces (such as at Rainbow Falls), in burrows on banks near seepages (Huheey 1964) and sometimes wandering through leaf litter on land, even during the day. Although considered an unusual behavior for this species, Jeff Corser (pers. comm.) spotted two Black-bellies on a tree trunk 90 cm (35.4 in) and 4.6 m (15 ft) above the ground surface, respectively, in Greenbrier. They often sit exposed on rocks within the current or along the stream margins, with the head, forepart of the body, or virtually the entire body out of water. For example, on 11 September 2000, my wife and I counted dozens of large adults fully exposed as we hiked along Alum Cave Creek; some almost appeared to be basking. Animals are active both diurnally and at night.

Gravid females have been observed on 9 February (1999), 2 May (1950), and 8 June (2000) in the Smokies, but oviposition usually occurs in May and June. Hatching occurs from July to September. For example, my wife and I uncovered a nest under a large rock in fast-flowing water at a small check dam in Fighting Creek on 5 August 2000. The embryos were in an advanced stage of development and, as the nest was uncovered, they hatched into the swift current. Most larvae remain in the larval condition for 3 to 4 years, and transform from late summer into autumn. Males mature at 6 years of age, and females at 7; these salamanders probably live into their early teens.

Black-bellied Salamanders eat virtually anything that they can catch, from small invertebrates to large salamanders. USGS survey crews observed them eating other salamanders *(D. ocoee,* smaller *D. quadramaculatus, D. santeetlah)* and crayfish (an 85 mm SVL salamander eating a 15 mm crayfish tail first; Palmer Creek, 17 July 2000). Cannibalism is not uncommon. In turn, they are preyed upon by reptiles, birds, and mammals; on 6 July 2000, we observed a 25 mm SVL Black-belly grasped firmly in the claws of a medium-sized egg-brooding crayfish.

Abundance and Status

Black-bellied Salamanders are abundant in most stream habitats throughout the Smokies. However, they are adversely affected by low pH. When the Anakeesta Formation was exposed during construction of the New-found Gap Road, this resulted in chemical interactions with the pyritic rock, which lowered the pH in surrounding streams. Larval Black-bellied (and other) Salamanders were eliminated from the road cut up to 1.6 km or greater downstream in Beech Flats Creek, and populations remained depauperate 20 years after contamination (Kuchen et al. 1994). The absence of salamanders at Icewater Spring along the Appalachian Trail east of Newfound Gap also may result from low pH as a result of contacts with the Anakeesta.

Remarks

Many Black-bellied Salamanders in the Great Smokies are found missing all or parts of their tails or limbs; many sport body scars, and one was captured during our surveys with a broken leg. Such injuries undoubtedly result from intraspecific aggression among the adults, as well as from predation attempts by large animals toward smaller animals, other salamanders, and avian and mammalian predators. They are very aggressive and will bite strongly when handled, although they are extremely difficult to grasp. Truly, they are the tigers of Southern Appalachian streams.

Santeetlah Dusky Salamander
Desmognathus santeetlah

Etymology

Santeetlah Salamander adult, Beech Flats at Luftee Gap.

Desmognathus: from the Greek *desmos,* meaning ligament, and *gnathos* meaning jaw. The name refers to the massive bundle of ligaments holding the jaw that gives a bulging appearance to the sides of the head of dusky salamanders; *santeetlah:* refers to several geographic localities of the same name in the Unicoi Mountains of the Southern Appalachians, and perhaps means "blue water" in the language of the Cherokees.

Identification

Adults. The Santeetlah Salamander is a medium-sized desmognathine sala-mander (see *D. imitator* account) that lives along streamsides. It is light to dark brown with a light belly; some individuals have a faintly reticulated or somewhat striped pattern, whereas others have no pattern at all. There is usually a light yellow wash on the undersides along the base of the tail. Older individuals may become melanistic, and thus be difficult to distin-guish from other medium-sized *Desmognathus.* Juveniles are brown with light bellies, and have 4 or 5 chestnut-colored spots between the limbs when viewed dorsally. As the salamander grows, these fuse to form the dorsal pattern. The tails are moderately keeled. Adults measure from 30 to 60 mm (1.2 to 2.4 in) SVL (Tilley 1981).

Santeetlah Salamander adult female guarding eggs. Kephart Prong (August 1999).

Santeetlah Salamander
(*Desmognathus santeetlah*).

Larvae. The larvae are small and brown, and have 5–8 alternating pairs of spots between the limbs when viewed from above. Spots con-tinue onto the tail.

Eggs. In the Great Smoky Moun-tains, female Santeetlah Salamanders deposit an average of 17–20 eggs (Tilley 1988). Large females lay more eggs than small females.

Similar Species. In the areas of con-tact or sympatry, uniformly dark-colored Imitator Salamanders may be easily confused with older Santeetlah Salamanders *(D. santeetlah).* When attempting to identify a medium-sized dusky salamander, read the var-ious species descriptions and look at the illustrations for comparison.

Santeetlah Salamanders usually have a light yellow wash on the underside of the limbs and tail (Imitators and Spotted Duskies do not). Imitators also have dark or black bellies. The pattern of Spotted Duskies is much

Eggs. Mature eggs are 2.5–3 mm in diameter, and 8–14 eggs are deposited singly (based on observations in Alabama). No data are available for the Great Smoky Mountains.

Similar Species. This species might be confused with the Long-tailed Salamander, although the color patterns of adults are distinctive. Both have very long tails with vertical bars on them. The larvae are very similar, however, but since the ranges of the species do not overlap, the larvae can be identified based on the location where they are observed: *E. longicauda* in Tennessee and *E. guttolineata* in North Carolina. Recently hatched larvae resemble the larvae of the Blue Ridge Two-lined Salamander, but they lack the paired light dorsal spot pattern of *E. wilderae*.

Taxonomic Comments. Until recently, the Three-lined Salamander was considered a subspecies of the Long-tailed Salamander, *Eurycea longicauda*. Although hybrids between these species have been reported elsewhere, the species are easily separable in the Great Smokies and no hybridization has been observed.

Distribution

Three-lined Salamanders are found generally on the coastal plain from northern Virginia south to the Gulf Coast, and north to western Tennessee. They occur at low elevations when they are found in the mountains. In the Great Smokies, this species is found only on the North Carolina side and is known from three recent records: in Big Cove near the Blue Ridge Parkway (on 8 July 2000) and from the vicinity of Glady Branch north of Fontana Reservoir (on 10 August 2000 and 8 May 2001). There are records for three additional specimens in the park's record books: two from 1.7 km (1 mi) west of Cherokee and one from Topton. However, these specimens could not be located in the park's collection and their whereabouts are unknown. Huheey and Stupka (1967) provided additional localities based on animals seen between 1949 and 1964, including on U.S. 441 1.6 km (one mile) south of Smokemont, near the Oconaluftee Visitor Center, near the confluence of Indian and Deep Creeks north of Bryson City, and from Cataloochee Creek. During fieldwork from 1998 to 2001, each of these locations, as well as Dunn's (1926) vaguely delineated collection site at "Mt. Sterling," were sampled on several occasions without success (see *E. longicauda* species account).

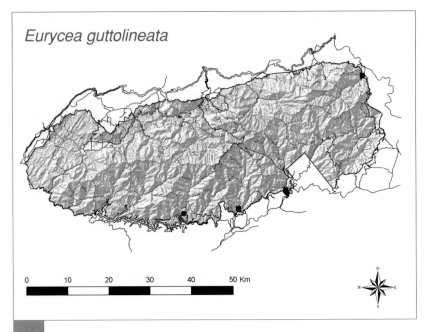

Eurycea guttolineata

Distribution of the Three-lined Salamander *(Eurycea guttolineata)*. The lower center square marks a historic locality at the junction of Indian and Deep Creeks; the square on the upper right indicates the "Mt. Sterling" location reported by E. R. Dunn.

Life History

Nothing is known specifically about the life history of this species within the Smokies. It appears to be associated with slow-moving or sluggish low-elevation streams. In other regions within its range, eggs are deposited in winter. The larval period is less than 1 year (ca. 3.5 to 5.5 months), and metamorphosis occurs during the summer. These salamanders eat a wide variety of invertebrates.

Abundance and Status

This species is rare within the Great Smokies, at least based on recent sightings. The species likely occurs more commonly than collections indicate in scattered lowland areas north of Fontana Reservoir east to and including the Cherokee Indian Reservation.

Remarks

Unlike many other species of lungless salamanders, this species does not defend territories.

Junaluska Salamander
Eurycea junaluska

Junaluska Salamander adult, Little River.

Etymology

Eurycea: the origin of this name is unknown. Its originator, Constantine Rafinesque, did not provide an etymology when he first used it in 1832; *junaluska:* name given in honor of the Cherokee Chief Junaluska.

Identification

Junaluska Salamander
(Eurycea junaluska).

Adults. The adults are medium-sized, with long limbs, a proportionally short tail, and a yellow-green to yellow-orange to cream-colored dorsum. Black flecks extend along the sides of the body and tail, onto the dorsum, although not in the same profusion. The belly is light and unpigmented. Adults measure 7.5–10 cm (3–4 in) TL.

Larvae. The larvae are very similar to the larvae of the Blue Ridge Two-lined Salamander, but are darker on the sides and tail, and have more prominent reticulations (Ryan 1997). Cheek patches are dense and well pigmented. The overall body color is olive green to light brown. These larvae also grow much larger (more than 42 mm [1.7 in] SVL) than the larvae of *E. wilderae.*

Eggs. Eggs average 3 mm in diameter, and the female deposits between 30 and 49 eggs per clutch. Eggs are attached singly or in small clusters under rocks in stream channels (Petranka 1998). Females brood their eggs.

Similar Species. The Junaluska Salamander most closely resembles the Blue Ridge Two-lined Salamander. However, it lacks the prominent black dorso-lateral stripe of *E. wilderae* and is more robust than that species. Blue Ridge Two-lined Salamanders usually are brighter with more distinctive, bold coloring than Junaluska Salamanders. In the smaller larval size classes, they may be impossible to distinguish from the much more common Blue Ridge Two-lined Salamander. However, the dorso-lateral contact between the dark sides and lighter dorsum on larvae is straight in *E. wilderae* and wavy in *E. junaluska*.

Taxonomic Comments. This enigmatic salamander often has been considered an aberrant *Eurycea wilderae*. However, molecular, morphological, and ecological data all suggest that it is a valid species.

Distribution

The Junaluska Salamander occurs only in extreme western North Carolina (Cheoah River drainage) and adjacent Tennessee (Tellico, Little Tennessee Rivers, West Prong of the Little Pigeon, and Little Pigeon River drainages).

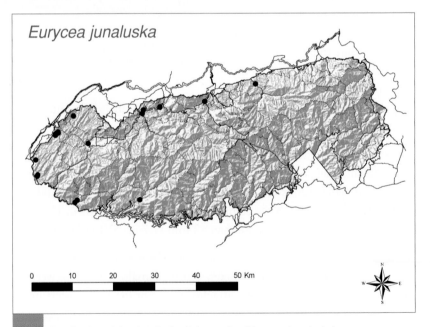

Distribution of the Junaluska Salamander *(Eurycea junaluska).*

In the Great Smoky Mountains, this species occurs in the Little Pigeon River in Greenbrier, lower Fighting Creek, Little River, Abrams Creek, Panther Creek, Tabcat Creek (and Cattail Branch), Twentymile Creek, and lower Hazel Creek. All these locations are lower than 524 m (1,720 ft) in elevation.

Life History

Little is known about the life history of this species within the park. Adults are very difficult to find until late summer and autumn in the Smokies, when they can be located under rocks and other surface objects along the margins of the larger streams. In other areas, they have been reported to be more active in mid- to late spring when egg laying occurs; females brooding eggs have been found in early May (Ryan 1998). Mating probably occurs in an extended period from autumn into the spring, as it does in some other park *Eurycea.*

Larvae are in the streams throughout the year (they take 1–3 years to metamorphose; see Bruce 1982; Ryan 1998), but small Junaluska Salamander larvae are very difficult to differentiate from larval *E. wilderae.* Larvae appear from the late spring to June. Second-year larvae are much larger than the largest larvae of *E. wilderae;* Jeff Corser found one measuring 47.7 mm TL on 4 April 2001 in Cane Creek. Larvae are conspicuous in August and September in certain streams of the Great Smokies, especially in the forming leaf mats of larger streams, such as in Abrams Creek in the western portion of Cades Cove. Jeff Corser observed individuals metamorphosing on 6 August (1999) in Tabcat Creek and at Metcalf Bottoms along Little River, but in other parts of its range metamorphosing Junaluska Salamanders have been reported in May (Ryan 1998). Metamorphosis normally occurs at 35 to 45 mm (1.4 to 1.8 in) TL.

Abundance and Status

Assessing the abundance and status of this species within the Great Smoky Mountains is hampered by the difficulty in finding these salamanders during much of the year. In some areas the species seems reasonably common at certain times (e.g., in autumn in Abrams Creek), but then it disappears for much of the year. Nothing is currently known about how Junaluska Salamanders could be affected by stream modification or pollution.

Remarks

The larvae of *E. wilderae* and *E. junaluska* often occur sympatrically, although larval *E. junaluska* are much less abundant than larval *E. wilderae.* How these larvae partition their environment is unknown, but Sever (1983) suggested that competition might be the reason that *E. junaluska* is so scarce.

Long-tailed Salamander
Eurycea longicauda

Etymology

Eurycea: the origin of this name is unknown. Its originator, Constantine Rafinesque, did not provide an etymology when he first used it in 1832; *longicauda:* from the Latin *longus,* meaning long, and *cauda,* meaning tail. The name refers to the long tail of this species.

Long-tailed Salamander adult, Abrams Creek.

Identification

Adults. The Long-tailed Salamander is a slender salamander with relatively large limbs and a very long tail (comprising more than 60 percent of the total body length). Adults measure 10–20 cm (4–8 in) TL. The background color is yellow to orange, with small dark spots dorsally and laterally. There is a series of parallel, vertical dark lines on either side of the tail. The belly is yellowish to cream color, usually without mottling. The eyes are prominent. Males have prominent cirri, whereas the cirri of females are not so obvious. Juveniles resemble the adults.

Larvae. Larvae are slender and streamlined for life in flowing water. They are uniformly darkly pigmented dorsally, with small unpigmented rounded spots on the upper side, with a second row of spots between the limbs. As the larvae get older, a light band forms dorsally, and it is suffused with black spots. Dark vertical, parallel bars on the tail are present in older larvae.

Eggs. Eggs measure 2.5 to 3 mm in diameter, and incubate 4–12 weeks depending on water temperature. Gravid females contain 60 to 100 eggs, but they probably lay their eggs in scattered locations underground, either singly or in small groups, rather than as a single clutch.

Long-tailed Salamander larva, Gregory's Cave (July 10, 2000).

Long-tailed Salamander *(Eurycea longicauda).*

Similar Species. This species most closely resembles the Cave Salamander, with which it is sympatric in some caves within the Great Smoky Mountains. Cave Salamanders are slightly more brightly colored, and do not have the vertical, parallel, dark bars on the tail. *Eurycea longicauda* might be confused with the Three-lined Salamander, although the color patterns of adults are distinctive. Both have very long tails with vertical bars on them. The larvae are very similar, however, but since the ranges of the species do not overlap, the larvae can be identified based on the location where they are observed: *E. longicauda* in Tennessee and *E. guttolineata* in North Carolina. Recently hatched larvae also somewhat resemble the larvae of the Blue Ridge Two-lined Salamander, but they lack the paired light dorsal spot pattern of *E. wilderae*.

Taxonomic Comments. Until 1997, *E. longicauda* was considered to have three subspecies: *longicauda, guttolineata,* and *melanopleura*. However, recent evidence suggests that *E. guttolineata* is sufficiently distinct to merit recognition as a full species (see *E. guttolineata* account). The subspecies of Long-tailed Salamander that occurs in the Great Smokies is *E. l. longicauda*.

Distribution

The Long-Tailed Salamander is found west of the Blue Ridge and Great Smoky Mountains, from southern New York south to northern Alabama, and west to southern Illinois and eastern Missouri. It is the "long-tailed" salamander of the Alleghenies, Cumberland Plateau, and rolling hills of southern Indiana and Ohio. In the Great Smokies, Long-tailed Salamanders are found only on the Tennessee side of the park. They occur at lower elevations west of Fighting Creek Gap in the Little River, Middle Prong of the Little River, Cades Cove, and Abrams Creek drainages. Specific locations include The Sinks, Whiteoak Sink, Gregory's Cave in Cades Cove, the Calf caves on Rich Mountain, the bottomlands along Cane and Beard Cane Creeks, and along creeks draining into Chilhowee Lake at the extreme west side of the park. Huheey and Stupka (1967) also recorded it from the Sugarlands and nearby Gatlinburg. In addition, Long-tailed Salamanders occur at Stupka's Cave on the Foothills Parkway right-of-way in Wear's

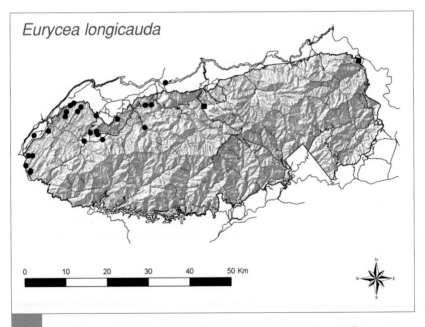

Eurycea longicauda

0   10   20   30   40   50 Km

Distribution of the Long-tailed Salamander *(Eurycea longicauda)*. The center square shows the historic location at Sugarlands; the square on the upper right shows the "Mt. Sterling" location reported by E. R. Dunn.

Cove. At night, they are found commonly on the rock faces along Laurel Creek Road between The Sinks and highway 73 into Townsend.

Dunn (1926) mentions a single Long-tailed Salamander collected at Mount Sterling, North Carolina, on the park's extreme eastern side. Surveys conducted between 1998 and 2001 at many potential sites in the vicinity of Big Creek and Mount Sterling have been unsuccessful in locating additional specimens. The exact location where the single specimen was taken is unknown, and the habitat has changed substantially since 24 July 1919 when the specimen was collected. It is interesting to note that all the other "long-tailed" salamanders preserved by Dunn from this location were collected on 25–26 July 1919 and were *E. guttolineata*. Both low- and high-elevation species (e.g., *Desmognathus monticola* and *D. wrighti*) were recorded with the same "Mt Sterling" location. As such, it seems unlikely that all specimens were collected from the immediate vicinity of Mount Sterling post office, a low-elevation site; the community of Mount Sterling more likely served as a focal point for collections throughout the Big Creek/Mount Sterling region. Dunn's collection is at the Museum of Comparative Zoology at Harvard.

Life History

Egg laying probably occurs in underground cracks and fissures from the autumn into the spring, with hatching occurring in March to April. Tiny larvae (10–15 mm TL) occur in Gregory's Cave by early April and persist until September, growing exceedingly slowly in this nutrient-poor environment. Few actually transform within the cave, as the pools usually evaporate before the larvae are large enough to transform (Dodd, Griffey, and Corser 2001). Larvae transformed only once in three years (7 September 2000), and then they only barely made it. On the surface, large larvae occur in small pools in Cades Cove by early to mid-June, and recently transformed juveniles have been seen at The Sinks by 4 July (1998). The larval period lasts less than one year, and larvae probably live in leaf mats. We also have captured larval Long-tailed Salamanders in clear, grassy pools in Cades Cove. Maturity occurs two years after metamorphosis. These salamanders may undertake migrations between summer and winter habitats and to breeding streams. Flies, beetles, and moths are favored prey, although a wide variety of invertebrates are eaten.

Abundance and Status

Long-tailed Salamanders appear reasonably common within the Great Smokies, and would probably seem more common if cliff faces and rock walls were searched at night. Nothing is known about how water quality affects survival.

Remarks

In the Great Smokies, both adults and juveniles of this species often are found near and within caves. At Gregory's Cave in Cades Cove, this species is found around the entrance during nearly all the warmer months, and larvae have been seen in small rimstone pools 73 m into the cave. Adults occur to 87 m from the entrance, well away from the twilight zone. The Long-tailed Salamander is not restricted to caves, as we have seen it terrestrially in many locations, always near streams or within old streambeds. At night, individuals can be seen along the moist rock faces at the Abrams Creek campground and along Little River Road.

Cave Salamander
Eurycea lucifuga

Etymology

Eurycea: the origin of this name is unknown. Its originator, Constantine Rafinesque, did not provide an etymology when he first used it in 1832; *lucifuga:* from the Latin *lucis,* meaning light, and *fuga,* meaning to flee. The scientific name refers to the cave environment this species prefers where it can "avoid light."

Identification

Adults. This is a long, bright red or orange salamander with dark spots scattered along its dorsum. The body is slender, the legs are proportionally large, and the tail is long (more than 60 percent of the total length) and lacks vertical, parallel, black bars; the eyes are conspicuous, the head is broad and flat, and the eyelids are gold flecked. The belly is white or cream-colored and unmarked. The tails are prehensile and are used to

Eggs. Eggs are white to pale yellow, and measure 2.5 to 3 mm. From ca. 20 to 100 eggs are deposited under rocks and other surface debris in small rivulets, seeps, and along streambanks.

Blue Ridge Two-lined Salamander (*Eurycea wilderae*).

Similar Species. The Junaluska Salamander most closely resembles the Blue Ridge Two-lined Salamander. However, it lacks the prominent black dorso-lateral stripe of *E. wilderae*, and is more robust than that species. Blue Ridge Two-lined Salamanders usually are brighter with more distinctive and bold coloring than Junaluska Salamanders. In the smaller larval size classes, Blue Ridge Two-lined Salamanders may be impossible to distinguish from the rare Junaluska Salamander. However, the dorso-lateral contact between the dark sides and lighter dorsum on larvae is normally straight in *E. wilderae* and wavy in *E. junaluska*. The small larvae also may be confused with the smaller larvae of the dusky salamanders. Viewed from above, the heads are more squarish than sympatric *Desmognathus* larvae, whose heads are rounded. *Desmognathus* larvae also have white gills, whereas the gills of *Eurycea* larvae are red.

Taxonomic Comments. *Eurycea wilderae* was originally described as a subspecies of the Two-lined Salamander, *E. bislineata.* It was elevated to specific status by Jacobs (1987). Petranka (1998) continued to recognize it as *E. b. wilderae,* noting that additional research on the contact zones between subspecies would be necessary, in his opinion, to validate its status as a distinct species. Although closely related, recent molecular studies carried out across such a contact zone provide evidence that the lowland form of the Bislineata complex, *E. cirrigera,* is indeed distinct from the highland *E. wilderae* (Kozak and Montanucci 2001). Bernardo (pers. comm.) also has been looking at molecular variation in *E. wilderae* within the Great Smokies; his preliminary data suggest that more than one species of "Blue Ridge Two-lined Salamander" may occur within the park. Further studies are planned.

Distribution

The Blue Ridge Two-line Salamander occurs in the Southern Appalachians from southwestern Virginia to northern Georgia. In the Great Smoky Mountains, this species is found throughout the park, from low elevations in the east (e.g., 335 m [1,100 ft] at the Abrams Creek Ranger Station) to

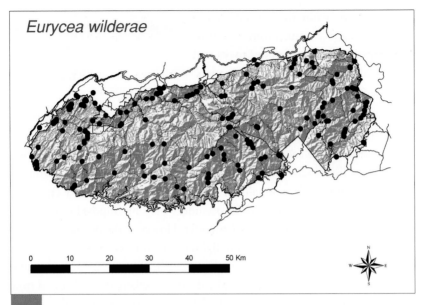

Distribution of the Blue Ridge Two-lined Salamander *(Eurycea wilderae)*.

1,783 m (5,850 ft) along Forney Ridge Trail on Clingmans Dome. It probably occurs at all elevations, even to the highest peaks.

Life History

In the Great Smokies, we have observed gravid females from 22 April to 22 September. Eggs are deposited in seeps and along streams under moss and other surface objects, and are guarded by the female. Eggs hatch between May and late September. However, numbers of gravid females are found in the autumn, and a clutch of 15 eggs with attendant female was observed on 2 February along the upper reaches of Falls Branch. It is unknown what stage of development the eggs were in, but these findings suggest that some females may brood through the winter, with the larvae hatching in spring. Large numbers of recently transformed larvae were seen on 5 June (1999) at Big Creek.

According to Petranka (1998), the length of the larval period and size at metamorphosis is highly variable in this species. In general, the larval period varies between 1 and 2 years; 1-year larvae metamorphose at 18 mm (0.7 in) TL, whereas 2-year larvae metamorphose at 32 mm (1.25 in) TL. These salamanders reach maturity at 3 to 4 years. Larvae forage continuously on aquatic crustaceans and zooplankton. Adults eat a wide variety of terrestrial invertebrates.

Abundance and Status

This is a very abundant salamander throughout the park. Indeed, the larvae of Blue Ridge Two-lined Salamanders may be the most abundant salamander species in the larger streams of the Great Smokies. In the Great Craggy Mountains of North Carolina, Petranka and Murray (2001) estimated that there were 1,490 Blue Ridge Two-lined Salamanders per hectare (2.2 acres).

Remarks

These species are not uncommonly found walking on the surface litter during daylight hours. Most activity likely occurs at night, and Blue Ridge Two-lined Salamanders often climb trees to forage. We even observed one adult in Abrams Creek under a flat rock in 17 cm (6.7 in) of water on 10 September (2000).

Spring Salamander
Gyrinophilus porphyriticus

Etymology

Gyrinophilus: from the Greek *gyrinos,* meaning "tadpole," and *phil,* meaning loving or fond of. Literally, fond of being a tadpole, perhaps referring to its extended larval stage; *porphyriticus:* from the Greek *porphyros,* referring to the reddish brown or purplish color of some specimens.

Spring Salamander adult, Andrews Bald. Note the black bordered light line, the canthus rostralis, which runs from the eye to the nostril.

Identification

Adults. Adult Spring Salamanders are large yellow-orange to salmon-colored salamanders. The dorsum has small dark flecks or spots; the flecks do not extend to the belly. Adults measure 11–21 cm (4.3 to 8.25 in) TL. Juveniles resemble the adults.

Spring Salamander larva, with growth caused by *Dermocystidium*. Cane Creek.

Spring Salamander
(Gyrinophilus porphyriticus).

Larvae. Spring Salamander larvae are streamlined and are light gray, yellow-brown, or lavender in coloration. They have long, squarish heads, small eyes, and fine, dark reticulations on the side of the body. There are no markings dorsally, as there are in *Pseudotriton* larvae.

Eggs. Females attach their eggs in a single layer to the underside of rocks or debris in water; few nests are known, and none have been found from the Great Smokies. A female deposits ca. 16 to 106 eggs per nest, each measuring 3.5–4 mm in diameter and having three layers of jelly membranes surrounding each individual egg. Large females deposit larger number of eggs than small females.

Similar Species. The species may be confused with either species of *Pseudotriton* in the Great Smokies, especially some individuals of *P. montanus*. However, *Gyrinophilus* has a prominent light line bordered in black or gray, which extends from the eye to the nostril (termed a canthus rostralis). *Pseudotriton* lack this line. The canthus rostralis is thought to act as a "gunsight" by helping the animal focus on a prey item.

Taxonomic Comments. There are four recognized subspecies of Spring Salamanders. Of these, Petranka (1998) shows the range of the Blue Ridge Spring Salamander *(G. p. danielsi)* as occurring on the North Carolina side of the Great Smokies, and the Northern Spring Salamander *(G. p. porphyriticus)* as occurring on the Tennessee side. As he notes, however, intergrades are known to occur, and it may be difficult to assign an individual salamander to a particular subspecies. According to Huheey and Stupka (1967), the patterns of Spring Salamanders change with elevation within the Great Smokies, and Brandon (1966) showed that there could

be continuous variation from low to high elevations. Brandon (1966) provides a detailed description of color and pattern variation in the Great
Smoky Mountains.

During surveys from 1998 to 2001, USGS field crews saw a large
number of Spring Salamanders within the park. All of the animals seen
conformed to the description of *G. p. danielsi,* regardless of whether they
were in Tennessee or North Carolina. Based on subjective impressions and
a re-examination of photographs and specimens, Spring Salamanders at
higher elevations in the Smokies seem to have larger and bolder spots and
flecks than do Spring Salamanders at lower elevations. However, the differences are not great. It may be biologically pointless to try to assign a
subspecific name to the phenotypically variable Spring Salamanders within
the Great Smoky Mountains. In any case, they do not conform to descriptions of *G. p. porphyriticus.*

Distribution

Spring Salamanders occur in the mountainous regions of Maine and
southern Québec south to the Piedmont Fall Line in South Carolina,

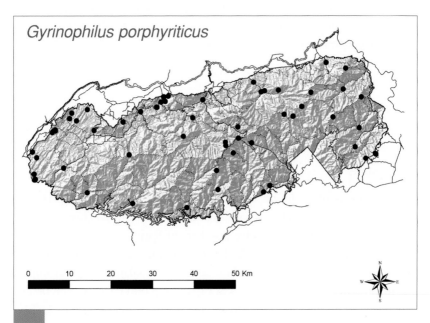

Distribution of the Spring Salamander *(Gyrinophilus porphyriticus).*

Georgia, and Alabama. In the Great Smoky Mountains, they are found throughout the park, essentially from the lowest areas around Chilhowee Lake to the top of Clingmans Dome.

Life History

Eggs are deposited in summer and hatch from late summer to autumn. A very large gravid female was seen on 3 April 2000 along Beardcane Creek (Jeff Corser, pers. comm.), and small larvae have been observed by 6 July. Larvae of various sizes are found in all types of flowing waters, from seeps to large streams, and even including roadside ditches. They may prefer the gravel interstices of streams, but we also have caught them using artificial leaf litter bags. Larvae are probably more active at night, when they emerge from hiding places to forage, than during the day. Bruce (1980) estimated the larval period to last as long as 4 years, and larvae transform when they reach 55–70 mm (2.2–2.75 in) TL. Metamorphosis occurs from June to August. Spring Salamanders are highly predatory and have been reported to eat other salamanders as well as a wide variety of invertebrates. They sometimes forage well away from streams and seeps, where they are found under surface litter. It may even be a more voracious predator than the Black-bellied Salamander. In the nearby Nantahala Mountains of North Carolina, Spring Salamanders are active throughout the year (Petranka 1998).

Abundance and Status

Spring Salamanders appear to be reasonably common within the Great Smokies, and no threats are currently known.

Remarks

This species is often found out walking on the surface during daylight hours, but it is most likely more nocturnal than diurnal. King (1939b) related an observation of massive numbers of Spring Salamanders that he witnessed on 20 April 1937. As he was driving across the park on the Newfound Gap Road, he counted ca. 200 of these salamanders crawling on the road between 1,067 and 1,524 m (3,500 to 5,000 ft) between 8:00 and 9:00 P.M. The weather was rainy, foggy, and warm. Similar observations have not been recorded since, but it is likely that these salamanders normally are active under such ideal environmental conditions. Along Beech Flats Creek, a USGS field crew observed two Spring Salamanders in trees well off the ground on 27 June 2000 (Barichivich, Smith, and Waldron 2001): a 64 mm (2.5 in) SVL animal 82 cm (32 in) above the ground surface, and a 68 mm (2.7 in) SVL animal 15 cm (5.9 in) above the ground surface.

One larva collected in Cane Creek by Jeff Corser on 30 March 1999 had a hyperplastic growth on its head; it was found to result from a fungus-like organism named *Dermocystidium.*

Four-toed Salamander
Hemidactylium scutatum

Etymology

Hemidactylium: from the Latin *hemi-* or Greek *hēmi,* meaning half, and the Greek *daktylion,* a term for the fusion of digits (fingers and toes). The reference presumably alludes to the absence of a fifth toe on each of the hind feet; *scutatum:* from the Latin *scutatus,* meaning covered with shieldlike plates. The reference may be to the appearance of the costal grooves on the side of the body or to the dorsal herringbone pattern, both of which resemble plates.

Spring Salamander adult, Andrews Bald. Note the black bordered light line, the canthus rostralis, which runs from the eye to the nostril.

Identification

Adults. Adults are small, slender salamanders with an obvious constriction at the base of the tail. The dorsal color is brown to reddish with a chevron or herringbone pattern, and the belly is bright white with conspicuous black spots. All feet have

Four-toed Salamander adult, showing brightly colored spotted belly. Cades Cove.

four toes, in contrast with other salamanders, which have four toes on their forefeet and five toes on their hind feet. Adults measure 5–10 cm (2–4 in) TL, and the juveniles resemble miniature adults.

Larvae. The larvae are slender and brown, with prominent eyes and a tail fin that extends the length of the body to near the head.

Eggs. Up to 80 white eggs are deposited by each female within a nest, but many females nest communally, making egg counts sometimes rather large per nest. Females may or may not guard their eggs; some females guard multiple clutches, whereas other nests are completely unguarded. The eggs measure 2.5–3 mm in diameter, and have a sticky coating which causes them to adhere to one another.

Four-toed Salamander
(Hemidactylium scutatum).

Similar Species. The combination of the tail constriction and the strikingly pigmented belly of this species distinguish it from all other salamanders in eastern North America. The only other Great Smokies salamander with four toes on the rear feet is the large and totally aquatic *Necturus maculosus*.

Taxonomic Comments. Despite the large geographic range and the often isolated distribution of populations, no subspecies are recognized.

Distribution

The Four-toed Salamander occurs in widely disjunct populations from Nova Scotia and southern Ontario to Louisiana and the Florida panhandle, and west to southeastern Oklahoma, central Missouri, and Wisconsin. In the Great Smokies, the Four-toed Salamander is found only on the Tennessee side of the park at lower elevations (less than 550 m [1,800 ft]). Locations include The Sinks, Cades Cove (Gum Swamp, bottomland-flooded forest along Abrams Creek), along Meadow Branch near Dorsey Gap (originally recorded 10 August 1941 and rediscovered by Jeff Corser on 5 April 2000), and in the Cane Creek and Beard Cane Creek drainages.

Life History

Eggs are laid early in spring in nests bordering still water, from small woodland pools to the large temporary pond in Gum Swamp. Although often associated with *Sphagnum,* this moss is not necessary for successful reproduction. We have seen nests from 7 March (1999) though 21 May (1999), with large numbers of brooding females from late March to mid-April. Eggs near hatching were observed by Jeff Corser on 14 May 1999 in pools along Cane Creek. Abandoned nests have been seen beginning in late March, suggesting that some females move to ponds in late January to February. Some females guard only their own egg clutches, some clutches are left unattended, and multiple clutches from different females some-times occupy a single nest. The larval period is very short, but nothing is known of its duration in the Smokies. In other populations, estimates range from 21 to 42 days. Sexual maturity occurs by age 2. In the non-reproductive season, adults are rarely encountered, although we have ob-served them from 8 April to 22 September in the Gum Swamp pond basin. Wintering aggregations of dozens of individuals have been found in Michigan, but whether similar aggregations occur in the Great Smoky Mountains is unknown.

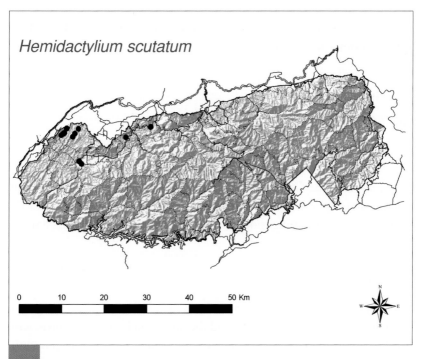

Hemidactylium scutatum

Distribution of the Four-toed Salamander *(Hemidactylium scutatum).*

Abundance and Status

This salamander is restricted to only a few specialized habitats within the Great Smokies, particularly in Cades Cove and in the Cane Creek drainage. During monitoring activities between 1999 and 2001, Jeff Corser (pers. comm.) counted between100 and 226 *H. scutatum* at nesting sites on the western side of the park; for an area the size of the Great Smoky Mountains National Park, this is not a large number of salamanders. These populations probably are very sensitive to disturbance, especially during the breeding season when the females brood egg clutches. Anything that alters water chemistry at Four-toed Salamander breeding sites would have a negative impact on their small, localized populations.

Remarks

The constriction at the base of the tail of this species may facilitate autotomy, but tail breakage does not appear to be more prevalent in this species than in any other terrestrial salamander. Perhaps this is because the tail is an important storage location for lipids, which are used by the females to provide yolk to their eggs and by all adults as a metabolic food source during the cold and dry seasons.

Common Mudpuppy
Necturus maculosus

Etymology

Necturus: from the Greek *nektos,* meaning swimming, and *oura,* meaning tail, presumably in reference to its aquatic existence; *maculosus:* from the Latin *maculosus,* meaning spotted. The name refers to the spotted dorsum of many adults.

Identification

Adults. This is a large, fully aquatic salamander with conspicuous dark red, bushy gills. The head is squarish, and the body is light to dark brown, with black spots or blotches on the dorsum and sides. The belly is grayish in color. Tails are paddle-shaped, and they are marked similarly to the rest of the body. There are four toes on each of the hind feet. Adults measure 20–30 cm (7.9–12 in) TL, although in other parts of their range they grow much larger. Juveniles resemble the larvae, but attain adult color patterns at 13 to 15 cm (5 to 6 in) TL.

Common Mudpuppy adult, location unknown. Photo by Richard Bartlett.

Larvae. The larvae are dark brown dorsally, with a mid-line stripe bordered on both sides by light yellow bands. Another dark stripe is located on the sides of the body. The gills are much less obvious than they are in the adult.

Eggs. Females attach from 40 to more than 150 cream to light yellow colored eggs in nests under rocks and other debris in slow-flowing water, generally less than 0.5 m (20 in) deep. The eggs measure 5–6.5 mm in diameter, and are surrounded by three jelly layers. The female guards the nest until the eggs hatch.

Common Mudpuppy larva, West Virginia. Photo by Richard Bartlett.

Similar Species. The only other large aquatic salamander in the Great Smokies is the Hellbender, which lacks the external red bushy gills and grows much larger and bulkier than the Common Mudpuppy. The larvae cannot be confused with other sym-

Common Mudpuppy (Necturus maculaosus).

patric salamander larvae because of their striking yellow and black coloration, and their possession of only four toes on each hind foot.

Taxonomic Comments. Two subspecies are currently recognized. The Common Mudpuppy (N. m. maculosus) is the subspecies found in the Great Smoky Mountains.

Distribution

The Common Mudpuppy occurs from southern Canada (Manitoba to Québec) south to the Tennessee River drainage in northern Alabama, Mississippi, and Georgia. It is most prevalent west of the Alleghenies and in the midwestern Great Lakes region. Records for the Common Mudpuppy are somewhat rare in the Great Smoky Mountains. The species is known from the lower reaches of Abrams Creek; Huheey and Stupka (1967) give collection dates of 6 September 1937 (2 adults, 6 larvae), 31 March 1940 (2 larvae), and June 1957. Most of the salamanders captured were larvae. Two additional larvae were found by Jamie Barichivich and Kevin Smith in Cottee Hole on Abrams Creek on 28–29 July 2000. Common Mudpuppies have been collected in the Little River near the park boundary at Townsend by NPS fish shocking crews, approximately 2.5 km (1.5 mi) east of the junction of highway 73 and Little River Road (Nickerson, Krysko, and Owen 2002), and 1.6 km (1.0 mi) from the park boundary (Merkle and Kovacic 1974).

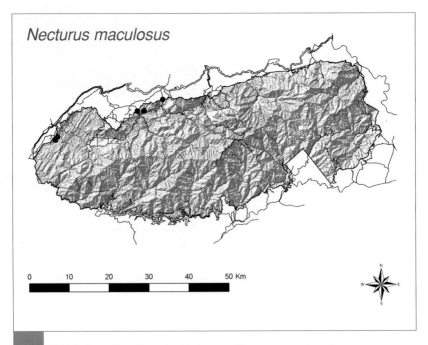

Distribution of the Common Mudpuppy *(Necturus maculosus)*.

Life History

Nothing is known concerning the life history of this species in the Great Smoky Mountains except for the collection dates. The animals captured at Cottee Hole were found in leaf litter piles within the creek. In other parts of its range, Common Mudpuppies deposit their eggs in May to early June, with hatching in July or August. Sexual maturity is reached at ca. five years of age. Common Mudpuppies eat a wide variety of prey and come out of their daytime hiding places to forage at night along stream bottoms. Unlike some other salamanders, Common Mudpuppies are active throughout the winter, where they can be found in the dense leaf mats that form in streams and rivers.

Abundance and Status

The Common Mudpuppy is considered relatively abundant throughout its range. In the Smokies, habitats are much more limited and the species seems somewhat rare. Perhaps this is because the streams offer only marginal habitat, or perhaps it is because so few biologists have actually looked for them within the park. This species also may have been adversely affected by native fish removal schemes undertaken in 1957 in connection with the introduction of rainbow trout. Abrams Creek, one of the streams treated with fish poison, still contains some Common Mudpuppies.

Remarks

The Common Mudpuppy is essentially a larval salamander that never undergoes metamorphosis. It resembles a larva and retains larval characteristics, but it is capable of breeding. Common Mudpuppies are therefore neotenic (i.e., they retain larval characteristics as an adult) and paedomorphic (i.e., they are sexually mature while retaining a larval body form). Although often used (incorrectly) interchangeably, these terms have slightly different meanings.

In some parts of its range, the Common Mudpuppy is an important host for a freshwater mussel, *Simpsonaias ambigua*. Larval salamander mussels have an immature stage, termed a glochidia, which is shed into the water. The glochidia attaches itself to the gills of the Common Mudpuppy and is transported and sheltered by the salamander. When it grows large enough, the glochidia detaches itself from the salamander, and settles onto the stream floor where it develops into an adult mussel. Unfortunately, the salamander mussel may be extirpated from Tennessee.

Eastern Red-spotted Newt adult, male in breeding condition. Methodist Church Pond in Cades Cove.

Eastern Red-spotted Newt adult, terrestrial phase. Gum Swamp.

Eastern Red-spotted Newt
Notophthalmus viridescens

Etymology

Notophthalmus: from the Greek *noto,* meaning a mark, and *ophtalmus* meaning eye, presumably in reference to the eye spots on the sides and back; *viridescens:* from the Latin *viridescens,* meaning slightly green. The name refers to the greenish color of the adults.

Identification

Adults. Terrestrial adults have a granular dry skin that is olive green with paired red spots dorsally, and a bright yellow belly with small black dots ventrally. Aquatic adults generally retain this coloration, but the skin becomes smooth and sleek, the ventral coloration is more subdued, and the tails become more paddle-shaped for aquatic locomotion. Males have a swollen vent and dark cornified toe tips and ridges on the inner thighs during the breeding season and a light orangish spot on the posterior part of the cloaca year round. Terrestrial efts are bright orange, with a granular skin, and are somewhat more compact than adults. They also have the bright red dorsal spots which are bordered in black.

Larvae. The larvae are small and colored greenish yellow dorsally and light yellow ventrally. There is an obvious separation between the dorsal and

Tennessee portion of Parsons Branch Road, many animals have a series of brassy dorsal flecks or spotting. Occasional individuals may be completely black or even more than 50 percent white. The bellies are dark blue to black and unspotted. Hatchlings and juveniles resemble small adults. The animal normally has 16 costal grooves. Adults measure 11.5–20.5 cm (4.5–8 in) TL.

Northern Slimy Salamander adult, Parsons Branch Road.

Larvae. This species has direct development; the young hatch as miniature adults.

Eggs. Females deposit egg clusters that are attached to the ceiling of the nesting cavity. Eggs are white and measure 3.5–5.5 mm in diameter. Usually ca. 5 to 15 eggs are found per clutch, but a female may not deposit her entire number of ovarian eggs within a single clutch. Females brood their eggs.

Northern Slimy Salamander adult, unusual color pattern. Ace Gap Trail.

Similar Species. This species most closely resembles the Southern Appalachian Salamander (which has fewer and smaller white spots or flecks on the back and sides, often is gray to slate in color, and has a lighter chin and belly). Undoubtedly, the Southern Appalachian Salamander has frequently been confused with the Northern Slimy Salamander in the older scientific literature (e.g., King 1939b; Huheey and Stupka 1967). It might also be confused with Jordan's Salamander (which has red cheeks) or the Gray-cheeked Salamander (which is uniform and lighter in color and lacks the white flecking and spots). The brassy spotted *P. glutinosus* on the western side of the Great Smokies closely resemble the published descriptions of the Tellico Salamander, *P. aureolus,* which occurs southwest of the Little Tennessee River.

Taxonomic Comments. The taxonomic status of the slimy salamanders in the Great Smoky Mountains is confusing. Within a radius of ca. 85 km (50 mi) of the park, at least five species of slimy salamanders (the Glutinosus complex) have been described: *P. aureolus, P. chattahoochee, P. cylindraceus, P. glutinosus, P. oconaluftee.* These species are differentiated on the basis of variation in gene frequencies among populations and differences in color pattern. Some authors (e.g., Frost and Hillis 1990; Petranka 1998) have chosen not to recognize each of these variants as separate species, arguing that genetic distance alone should not be used as a primary criterion to separate species. Also, described differences in color patterns often do not match animals in the field. In various areas within the Great Smoky Mountains, the color patterns described for the "species" listed above sometimes are seen in slimy salamanders within the same population (e.g., along Parson Branch Road at Weasel Branch).

The distribution maps in Petranka (1998) and Highton and Peabody (2000) imply that the Northern Slimy Salamander occurs mostly on the Tennessee side of the Park, with the Southern Appalachian Salamander occurring mostly on the North Carolina side. If this is the case, the salamanders in the field certainly do not match the morphological descriptions (Highton, 1989; Petranka, 1998; Highton and Peabody 2000) for these species. Both *P. glutinosus* and *P. oconaluftee* have essentially the same life history, eat the same foods, and occur in similar habitats. Given the disagreement among taxonomists and the variation in morphology, it is difficult to tell which species occurs at a particular site without referring to molecular data. Even then, differentiating between Northern Slimy Salamanders, Southern Appalachian Salamanders, and possibly other members of the Glutinosus complex, is often difficult, even for experienced researchers.

Distribution

Members of the slimy salamander complex occur from New Hampshire to central Florida and west to central Texas and eastern Oklahoma. Slimy salamanders are absent from the prairie regions of the Midwest. The Northern Slimy Salamander, *P. glutinosus,* occurs from New Hampshire south to eastern Alabama and northern Georgia, and west to southern Illinois and Indiana. In the Great Smoky Mountains, salamanders resembling the

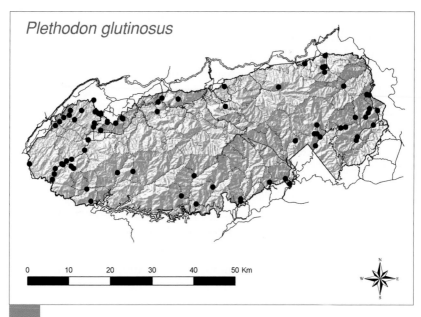

Plethodon glutinosus

| 0 | 10 | 20 | 30 | 40 | 50 Km |

Distribution of the Northern Slimy Salamander *(Plethodon glutinosus)*.

published descriptions of *P. glutinosus* are found throughout the park at mid- to lower elevations. During field surveys from 1998 to 2001, Northern Slimy Salamanders were found from 585 m (1,920 ft) along Forge Creek up to 1,280 m (4,200 ft) just west of Mount Sterling Gap on the Mount Sterling Trail. A large majority of the sites where this species was found were below 975 m (3,200 ft), however. I suggest that rather than replacing one another, the ranges of *P. glutinosus* and *P. oconaluftee* are broadly overlapping in the Smokies.

Life History

This species is entirely terrestrial, living under surface forest litter and in tunnels and crevices under the forest floor. Adults and juveniles are most often seen during the warmest weather from mid-April to early October, but some individuals can be seen at nearly any month of the year. Jeff Corser has recorded their surface presence in January in the Cane Creek region, a low-elevation site in the park's northwestern corner. Activity varies with elevation (later as elevation increases) and rainfall (they are not active in dry weather). Eggs presumably are deposited underground in late spring to early summer; very few nests have ever been located, and none apparently in the Smokies. Hatching occurs in the autumn, but some small animals may be found virtually at any time of the year, as it takes several years (probably 3–4) to reach reproductive maturity. These salamanders consume a wide variety of invertebrates, especially ants, beetles, and sowbugs.

Abundance and Status

The Northern Slimy Salamander appears to be relatively common throughout the park, especially on the Tennessee side of the Smokies. The most my wife and I ever captured during a single 30-minute timed survey between 1998 and 2001 was 13; in most surveys where this species was found, only 1 to 5 animals were normally encountered, however.

Remarks

These salamanders are most often encountered at night walking around on the surface litter, although they also are occasionally seen in daylight, especially during and after rains. They hide under rocks, logs, and other surface debris and may occasionally be encountered in rock crevices. Northern Slimy Salamanders readily are found around cave entrances virtually throughout the warm-weather season; from July to September, they move deeper into the caves (Dodd, Griffey, and Corser 2001). We have observed them as far as 40 m (132 ft) into Gregory's Cave in Cades Cove, although the average distance was 26.5 m (87 ft) based on 16 visits between 1998 and 2000.

As their name implies, Northern Slimy Salamanders are *slimy.* They give off a copious, white skin secretion, particularly on the tail, when handled or attacked by a predator. The secretion originates from granular glands in the skin, and it is noxious rather than toxic to would-be predators. The secretion tastes bad (yes, I have tasted it!), and it is sticky and gummy, much more so than that of any other *Plethodon,* and adheres to a predator's face, fur, or feathers. The secretion is difficult to remove, even by scrubbing, and may remain on one's hands for days after contact. It is not harmful to humans.

Jordan's Salamander
Plethodon jordani

Etymology

Jordan's Salamander adult,
Clingmans Dome.

Plethodon: from the Greek words *plethore,* meaning fullness or full of, and *odon,* meaning teeth; literally full of teeth. The name reflects the large number of teeth on both the jaws and the roof of the mouth (the vomerine teeth); *jordani:* named in honor of the ichthyologist David Starr Jordan (1851–1931) of Stanford University.

Identification

Adults. This is a medium-sized blue-black terrestrial salamander with bright red, occasionally orange or yellow, cheek patches. The bellies are gray to black. There are no white spots or red coloration anywhere on the dorsum, sides, or bellies. Hatchlings and juveniles resemble the adults. Costal grooves: normally 16. Adults measure 8.5–18.5 cm (3.3–7.3 in) TL.

Larvae. This species has direct development, and the young hatch as miniature adults.

Eggs. Nests of *P. jordani* have never been found in the wild. Apparently, no information on the eggs or clutch size has ever been published.

Similar Species. *Plethodon metcalfi* and *P. jordani* are virtually identical, except for their cheek coloration. Their ranges do not overlap, although hybridization occurs in contact zones in the Balsam Mountains. Hybrids usually have part red and part gray cheeks. Other salamanders with partial red and gray cheeks occur on the northern slopes of Mount Le Conte (along the Rainbow Falls and Maddron Bald Trails). In this case, hybridization most likely is occurring between *P. jordani* and *P. oconaluftee.* Other potential hybrids between these latter species were observed along the upper portions of Beech Flats Creek on 2 June 1999. Dawley (1987) mentions that hybrids between these latter species also occurs on the slopes of Mount Sterling.

Taxonomic Comments. Because of its overall restricted range, no subspecies are recognized. Jordan's Salamander hybridizes with the Oconaluftee Salamander and the Southern Gray-cheeked Salamander (see the accounts for these species) within the Great Smokies. Petranka (1998) did not recognize any of the other members of the Jordani complex as separate species. However, I follow Highton and Peabody (2000) in differentiating *P. jordani, sensu stricto*, from the six other species of salamanders formerly considered within the Jordani complex, including *P. metcalfi* in the Balsam Mountains within Great Smoky Mountains National Park. Highton and Peabody (2000) provide an extensive review of the nomenclature within the Jordani complex and molecular evidence validating their decision to partition the complex. Still, some researchers do not believe that partitioning of the species is warranted at this time.

Distribution

This species occurs only in the Great Smoky Mountains National Park. It is found from Mount Sterling Gap in the east to the slopes of Gregory Bald in the west. The lowest elevation that it was found during the 1998 to 2001 field surveys was 775 m (2,545 ft) in the Solola Valley on the North Carolina side of the park. The species occurs to the highest elevations of Clingmans Dome (2,025 m [6,643 ft]).

Life History

Jordan's Salamanders are entirely terrestrial, and are most commonly found from the spring to the autumn. They occur under bark, logs, rocks, and other surface debris on the forest floor and are frequently seen at night walking on the humid forest floor. This is a cool- and wet-weather-loving species; my wife and I even have collected it at Indian Gap on the Clingmans Dome road during a 23 April 1998 snowstorm. These salamanders feed on a wide variety of forest floor invertebrates (Powders and Tietjen 1974), and one was captured in the process of devouring a large worm on 7 June (2000) on Clingmans Dome. There does not appear to be any

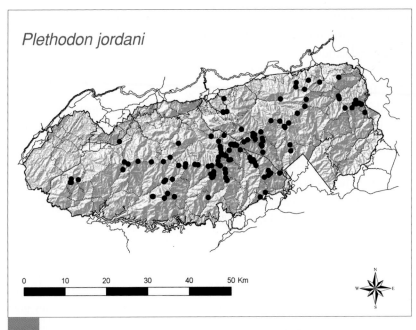

Distribution of Jordan's Salamander *(Plethodon jordani)*.

evidence of food competition between these and other sympatric wood-land salamanders within the Great Smokies. Females most likely deposit their eggs in early summer, with hatching occurring in late summer to early autumn. Small juveniles are most common late in the season, but juveniles can be found at all times of the activity period. For instance, we observed large numbers of tiny Jordan's Salamanders on Fork Ridge at the head of Keg Drive Branch on 17 May 1998, but these were most likely young from the previous autumn. Males mature in 3 years, and females mature in 4 to 6 years.

Abundance and Status

Jordan's Salamander is generally abundant within the Great Smokies. It was quite common to count more than 15 *P. jordani* during a 30-minute sampling period during field surveys between 1998 and 2001 and, on one occasion, my wife and I counted 67 *P. jordani* during a single sampling period (9 May 1999) on top of Mount Sterling. In the Great Smokies,

Jordan's Salamanders are found in significantly higher numbers on sites without a previous history of disturbance by farming or logging (Lydic 1999; Hyde and Simons 2001). Merchant (1972) estimated that males had a home range of 11.4 m^2 (123.5 ft^2), females had a home range of 2.8 m^2 (30.3 ft^2), and juveniles had a home range of 1.7 m^2 (18.5 ft^2). In contrast, Nishikawa (1990) estimated that male home ranges averaged 5.04 m^2, and female home ranges averaged 1.87 m^2. The home ranges of this species were not associated with cover objects or retreat holes.

Collectors occasionally illegally take these endemic salamanders from the park. These salamanders should only be enjoyed in the wild, as should all the park's amphibians, and no animal ever should be collected without permission from the National Park Service.

Remarks

Jordan's Salamanders can be found both at night and during the day. At night, they freely walk across the forest floor but, during the day, they generally seek cover. Some animals do leave their retreats during the day and can be seen walking on the surface litter, especially during very cloudy or rainy, cool days. Activity occurs during all the warmer months of the year at higher elevations and probably continues year-round as long as the ground does not freeze. Although primarily terrestrial and perhaps subterranean to some degree, Jordan's Salamanders occasionally climb trees. For example, we observed two adults climbing Fraser firs on 18 July 2000: one was 1 m (39.4 in) above the ground on the bark of an upright tree, whereas the other was 75 cm (29.5 in) above the surface under the bark of a fallen tree.

As with the other larger *Plethodon, P. jordani* gives off a copious white skin secretion, particularly on the tail, when handled or attacked by a predator. The secretion originates from granular glands in the skin and is noxious rather than toxic to would-be predators. The secretion tastes bad (I have tasted it!), and it is sticky and gummy, thus adhering to a predator's face, fur, or feathers. The secretion is difficult to remove, even by scrubbing. The red coloration of the cheeks may serve as a warning signal (termed an aposematic signal) to predators so that they learn to avoid this species of salamander. The quite edible Imitator Salamander mimics the noxious Jordan's Salamander and thus gains a survival advantage (Brodie and Howard 1973).

Any predator that had encountered a distasteful red-cheeked *P. jordani* would likely also avoid a red-cheeked *D. imitator*. This type of mimicry is termed Batesian mimicry.

In a series of intriguing experiments, Dawley (1987) examined the mating preferences of adult Jordan's Salamanders and Southern Appalachian Salamanders in an area where hybridization occurs between these species (Mount Sterling) and in an area where it does not (along Kephart Prong). Where no hybridization occurs, the males of each species preferred the odors of its own species to that of the other. However, where hybridization was occurring, only male *P. oconaluftee* were able to discriminate odors, that is, both sexes of *P. jordani* and female *P. oconaluftee* could not. This breakdown in ability to discriminate chemical odors could lead to the observed hybridization.

Southern Gray-cheeked Salamander
Plethodon metcalfi

Etymology

Southern Gray-cheeked Salamander adult, Heintooga Road in Balsam Mountains.

Plethodon: from the Greek words *plethore,* meaning fullness or full of, and *odon,* meaning teeth; literally full of teeth. The name reflects the large number of teeth on both the jaws and the roof of the mouth (the vomerine teeth); *metcalfi:* Latinized name in honor of North Carolina State University entomologist Zeno Payne Metcalf (1885–1956). In the 1920s, Metcalf led efforts in opposition to legislation that would have banned the teaching of evolution in North Carolina schools.

Identification

Adults. This is a medium-sized blue-gray to blue-black terrestrial salamander with light gray cheek patches. The bellies are gray. There are no white spots or red coloration anywhere on the dorsum, sides, or bellies. Hatchlings and juveniles resemble the adults. Costal grooves: normally 16. Adults measure 8.5–18.5 cm (3.3–7.3 in) TL.

Larvae. None. This species has direct development.

Eggs. Nests of *P. metcalfi* have never been found in the wild. Apparently, no information on the eggs or clutch size has ever been published.

Similar Species. *Plethodon metcalfi* and *P. jordani* are virtually identical, except for their cheek coloration. Their ranges do not overlap, although hybridization occurs in contact zones in the Balsam Mountains. Hybrids have part red and part gray cheeks.

Taxonomic Comments. According to Highton and Peabody (2000), *P. metcalfi* hybridizes with *P. jordani* on both Hyatt Ridge and Balsam Mountain. Hairston, Wiley, and Smith (1992) found that the width of the hybrid zone between these species did not change over an 18-year period. The presence of Jordan's Salamanders on Mount Sterling at 1,755 m (5,760 ft) with partial or full gray cheeks suggests that Southern Gray-cheeked Salamanders also hybridize with Jordan's Salamander at high elevations farther east of Balsam Mountain. In some of the older literature, *P. metcalfi* is listed as a subspecies of *P. jordani*. Petranka (1998) did not recognize any of the members of the Jordani complex as separate species from *P. jordani*. However, I follow Highton and Peabody (2000) in differentiating *P. metcalfi* from the six other species of salamanders formerly considered within the Jordani complex.

Distribution

The Southern Gray-cheeked Salamander is found only in western North Carolina and the adjacent mountains and upper Piedmont of South Carolina. In the Great Smoky Mountains, this species is found only on Hyatt Ridge, Balsam Mountain, and along Cataloochee Divide (Highton and Peabody, 2000; also see taxonomic comments). During field surveys in 1998 through 2001, the species was observed from 1,220 m (4,000 ft) along Rough Fork Trail on the north slope of Big Fork Ridge to 1,658 m (5,440 ft) near the Heintooga Overlook picnic area. It likely occurs to the highest parts of the Balsams (higher than 1,768 m [5,800 ft]).

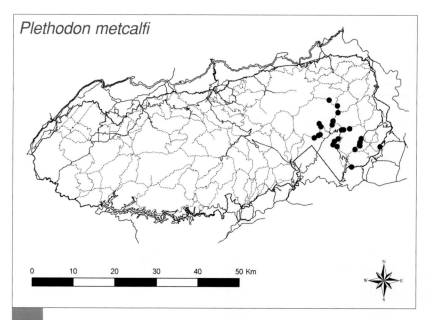

Plethodon metcalfi

0 10 20 30 40 50 Km

Distribution of the Southern Gray-cheeked Salamander *(Plethodon metcalfi)*.

Life History

The life history of this species in the Balsam Mountains is essentially the same as that of *P. jordani*. This species is active most conspicuously for 2 to 4 hours after dusk (Hairston 1987), but they can be found under leaf litter throughout the day and sometimes can be found walking through the leaf litter during the day. Interference competition, rather than competition for food, may play an important role in the spacing of terrestrial woodland salamanders on the forest floor.

Abundance and Status

This species appears to be fairly common in the Great Balsams. Hairston (1983) found that adults had an 81 percent survivorship and a mean generation time of 10 years. In count surveys over a period of 21 years, Hairston and Wiley (1993) detected no differences in the number of Gray-cheeked Salamanders observed at permanent study plots near the Heintooga Overlook. Although they stated that no declines had been observed, the relationship between the counts and actual population size is unknown. There is no indication, however, that this species faces any threats of population decline within the park.

Remarks

Much of the classical ecological work on salamander communities by Hairston (summarized in his 1987 book) actually refers to studies conducted on this species, rather than *P. jordani* as presently understood. Part of Hairston's work was conducted in the Balsam Mountains, primarily along the Balsam Mountain Road.

Southern Appalachian Salamander adult, along Balsam Mountain Road near Ledge Creek.

Southern Appalachian Salamander
Plethodon oconaluftee

Etymology

Plethodon: from the Greek words *plethore,* meaning fullness or full of, and *odon,* meaning teeth; literally full of teeth. The name reflects the large number of teeth on both the jaws and the roof of the mouth (the vomerine teeth); *oconaluftee:* the name Oconaluftee is an incorrect form of the Cherokee word egwanulti, which means "all towns upon the river" (Coggins 1999). As used with the salamander, the name refers to the Oconaluftee River in the Great Smoky Mountains.

Identification

Adults. This is a medium-sized, terrestrial salamander with a dorsal color of slate blue to nearly black. There are scattered tiny white flecks or spots on the dorsum and sides, although the sides may contain considerable white spots and flecking where they become conspicuous. The bellies have the same general ground color as the dorsum, usually some shade of gray,

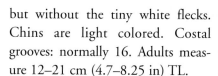

but without the tiny white flecks. Chins are light colored. Costal grooves: normally 16. Adults measure 12–21 cm (4.7–8.25 in) TL.

Larvae. This species has direct development; the young hatch as miniature adults.

Eggs. This species undoubtedly nests underground, but the eggs have not been discovered. Presumably they resemble the eggs of *P. glutinosus*.

Similar Species. This species most closely resembles Jordan's Salamander (but lacks the red cheeks), the Gray-cheeked Salamander (which is uniform in color and lacks the white flecking and spots), and the Northern Slimy Salamander (which has many more and larger white or brassy spots on the back and sides and has a dark chin and belly). Undoubtedly, the Southern Appalachian Salamander has frequently been confused with these species, especially the Northern Slimy Salamander, in the older scientific literature.

Southern Appalachian Salamander large juvenile, *P. oconaluftee × P. jordani* hybrid, along Roaring Fork Motor Trail. Individuals from this area have gray to red cheeks with numerous spots characteristic of *P.oconaluftee*.

Taxonomic Comments. There is some disagreement in the literature with regard to the correct scientific name of this species. It was originally described as *Plethodon teyahalee* from the Snowbird Mountains of North Carolina, but subsequent work demonstrated that these animals were actually hybrids. For this reason, Hairston (1993) suggested the name *P. oconaluftee* for the species elsewhere and reallocated the type locality to the Balsam Mountains of Transylvania County in North Carolina. Other biologists (Highton and Peabody 2000) have objected and point out that the rules of nomenclature regarding the naming of species apply only to certain types of hybrids and may not negate the name *teyahalee*. It seems likely that debate will continue but, for the purposes of this field guide, I follow the arguments in Petranka (1998) and use the name *P. oconaluftee*.

Regardless of the name used, the Southern Appalachian Salamander hybridizes with other large *Plethodon,* particularly *P. jordani* and

P. glutinosus, and Highton (1989) speculated that *P. oconaluftee* originated from hybridization between these two species. In the Great Smokies, *P. oconaluftee* appears to hybridize with *P. jordani* on the northeast slope of Maddron Bald (along Maddron Bald Trail) at an elevation of 865 m (2,840 ft) and on the north and at a similar elevation on the northwest slopes of Mount Le Conte and Mount Sterling; phenotypically pure populations of *P. oconaluftee* are found at lower elevations of Maddron Bald. Near the Baxter cabin on the Maddron Bald Trail (661 m [2,170 ft]) and along Rabbit Creek (near Weasel Branch) on the Parson's Branch Road (768 m [2,520 ft]), this species appears to coexist with brassy and profusely white-spotted *P. glutinosus* without hybridizing. See the *P. glutinosus* account for additional comments concerning the systematics of this species.

Distribution

The Southern Appalachian Salamander is known only from west of the French Broad River in southwestern North Carolina, a very small area of adjacent Tennessee, Rabun County in Georgia, and in extreme northwestern South Carolina. In Great Smoky Mountains National Park, the

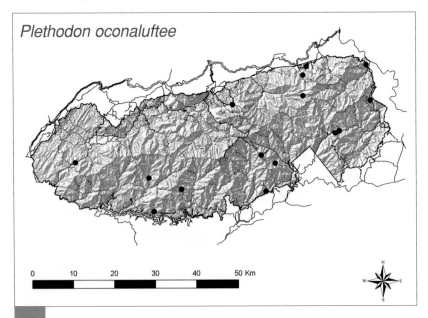

Distribution of the Southern Appalachian Salamander *(Plethodon oconaluftee).*

species is found at middle elevations throughout the park. In field surveys from 1998 to 2001, this species was recorded from Parson's Branch Road to the Balsam Mountains (just south of Pin Oak Gap) in North Carolina, and from the Roaring Fork Motor Trail east to the Maddron Bald Trail to Albright Grove. It was found from 649 m (2,130 ft) along the lower Maddron Bald Trail to 1,516 m (4,975 ft) along Thomas Ridge northwest of Nettle Creek Bald. Most locations are well below the elevation of the latter locality, however. The only place where the Southern Appalachian Salamander was found in numbers was along the Maddron Bald Trail over-looking Indian Camp Creek at elevations between 756 and 866 m (2,480 and 2,840 ft). Dawley (1987) reported a nonhybridizing population along Kephart Prong. Because of confusion in identifying this species, it proba-bly occurs more widely than currently indicated.

Life History

Little is known concerning the life history of this species within the Great Smoky Mountains. It probably has very similar habits and natural history to *P. jordani*. Breeding occurs from July to October, and eggs are probably deposited in the late spring to early summer. Hatching occurs after 2 or 3 months, with maturity occurring at 5 years of age, at least for females. The Southern Appalachian Salamander spends a significant amount of time in retreat holes, emerging at night to forage on the forest floor in the leaf litter. Like other terrestrial *Plethodon,* it eats a wide variety of forest floor invertebrates.

In a study of movement patterns in the Balsam Mountains, Nishikawa (1990) determined that adult Southern Appalachian Salamanders occu-pied defined home ranges, each with a primary retreat site. The size of the home range varied considerably, from 0.01 to 4.69 m^2, but males and females had similar home range sizes. Of more interest, perhaps, is that the home ranges of these salamanders tended not to overlap, suggesting that they are territorial.

Abundance and Status

Merchant (1972), presumably studying this species (he calls it *"Plethodon glutinosus"*), observed a density of 0.23 *P. oconaluftee* per m^2 (1 per 46 ft^2) at 1,219 m (4,000 ft) along Taywa Creek. We never observed large numbers of this species in samples collected from 1998 to 2001 throughout the park, except once. On 8 September 2000, our USGS field crew captured 19 adults and juveniles in a single 30-minute survey (see the distribution heading above). At other locations, only occasional individuals were observed, even on plots worked intensively over a 3-year period. There are no indications, however, of declines in this species.

Remarks

In a series of intriguing experiments, Dawley (1987) examined the mating preferences of adult Southern Appalachian Salamanders and Jordan's Salamanders in an area where hybridization occurs between these species (Mount Sterling) and in an area where it does not (along Kephart Prong). Where no hybridization occurs, the males of each species preferred the odors of its own species to that of the other. However, where hybridization was occurring, only male *P. oconaluftee* were able to discriminate odors, that is, both sexes of *P. jordani* and female *P. oconaluftee* could not. This breakdown in ability to discriminate chemical odors could lead to the observed hybridization. Why certain populations of salamanders can discriminate odors, whereas others cannot, remains a mystery. In any case, odors and chemical cues form a very important part of the life of forest floor terrestrial salamanders.

Southern Red-backed Salamander adults, red-striped phase (upper) and dark phase (lower), both from forest adjacent to Methodist Church Pond in Cades Cove.

Southern Red-backed Salamander
Plethodon serratus

Etymology

Plethodon: from the Greek words *plethore,* meaning fullness or full of, and *odon,* meaning teeth; literally full of teeth. The name reflects the large number of teeth on both the jaws and the roof of the mouth (the vomerine teeth); *serratus:* refers to the serrated dorsal red stripe that is characteristic mostly of the Louisiana and Ouachita Mountain isolates of this species. In the Smokies, the dorsal stripe is usually not (or only faintly) serrated.

Identification

Adults. This is a small woodland terrestrial salamander with a bright red, dorsal, straight-edged stripe. The stripe is usually red, but some animals may have yellowish to gold stripes. The sides of the body are dark gray to brown or black, and the belly is usually a mottled light to dark gray. White spots may be seen on the sides of some animals. Occasionally, individuals lack the red stripe and are uniformly dark colored. Such animals rarely occur in the Great Smokies, but they have been seen in the woodlands next to Methodist Church Pond in Cades Cove. Costal grooves: 18–21. Adults measure 6.5–10.5 cm (2.5–4.1 in) TL. The hatchlings and juveniles resemble tiny adults.

Larvae. There is no larval stage. Southern Red-backed Salamanders have direct development within the egg.

Eggs. Eggs have not been discovered in nature, but are probably deposited deep underground. In captivity, they deposit a mean number of 5 to 7 eggs per clutch; each egg measures 4.5 mm in diameter. Females brood their eggs.

Similar Species. The Southern Red-backed Salamander somewhat resembles the Seepage Salamander and some color patterns found in Spotted Dusky Salamanders and in Ocoee Salamanders. Southern Red-backed Salamanders are more terrestrial than Seepage Salamanders, and they do not possess the light line from the eye to the back of the jaw characteristic of all salamanders of the genus *Desmognathus.* The Ocoee Salamander is more robust than the Southern Red-backed Salamander, and their ranges in the Smokies do not overlap to any extent; the Ocoee Salamander is found only at the highest elevations. Small Spotted Dusky Salamanders occasionally have a straight red stripe, but like other *Desmognathus,* they have a light line from the eye to the back of the jaw. In the Southern Zigzag Salamander, the dorsal stripe is usually darker and zigzagged.

Taxonomic Comments. This species was once considered conspecific with the Red-backed Salamander, *P. cinereus,* and older literature on the amphibians of the Great Smokies (e.g., King 1939a, 1939b; Huheey and Stupka 1967) refer to it under this name. Molecular evidence indicates that they are really a separate species from their northern counterpart.

Distribution

The Southern Red-backed Salamander is found in scattered locations in the Southern Appalachians (southwestern North Carolina, adjacent Tennessee, northwestern Georgia), southeastern Missouri, west-central Arkansas and adjacent Oklahoma, and in central Louisiana. In the Great Smoky Mountains, it is found throughout the park, most commonly at middle to lower elevations. During the 1998 to 2001 field surveys, the Southern Red-backed Salamander was found from 360 m (1,180 ft) along Kingfisher Creek, a tributary to Abrams Creek on the park's far western side, to more than 1,524 m (5,000 ft) on Roundtop Knob (1,540 m [5,050 ft]) and Beetree Ridge (1,527 m [5,010 ft]). Huheey and Stupka (1967) recorded it from as high as Mollie Gap (1,685 m [5,530 ft]) on a narrow strip of the Heintooga Ridge Road connecting the Blue Ridge Parkway and the Balsam Mountains.

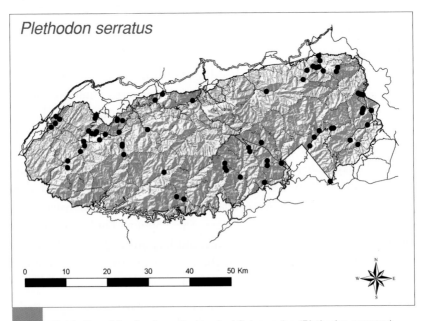

Distribution of the Southern Red-backed Salamander *(Plethodon serratus)*.

Life History

Southern Red-backed Salamanders are entirely terrestrial, living under rocks, logs, and surface debris. Mating occurs in the late winter to early spring; one gravid female was observed on 6 March 2000 on Rich Mountain Road at Hesse Creek (J. Corser, pers. comm.). The eggs are deposited in underground nests in June and July, and the young hatch in the late summer. Red-backed Salamanders of very small size were seen on 7 April 1999 in the vicinity of Hen Wallow Falls, but these animals actually may have hatched the previous autumn. *Plethodon serratus* probably takes about two years to reach sexual maturity. Southern Red-backed Salamanders may be active day or night, but they are most often found at night moving through surface leaf litter. They are most active in spring (March to May) and autumn (October to November) at lower elevations, but as elevation increases, they are more likely to be found during the summer. At lower elevations, they can be found throughout the winter on warmer days. For example, Jeff Corser found an individual near Cane Creek on 15 January 2001, and numerous individuals were seen in Gum Swamp on 19 January. Food consists of any small invertebrate; we observed one eating a small centipede in Cades Cove in April 2000.

Abundance and Status

The Southern Red-backed Salamander is a common species within the park. This species may be particularly sensitive to acidic soils, as is *P. cinereus*, but whether this sensitivity is reflected in their distribution within the park is unknown.

Remarks

A large amount of research has been conducted on the use of chemical cues by the closely related species, *P. cinereus*, which also may apply to *P. serratus*. It is known that individuals can identify one another, that males defend territories, and that females can even determine the fitness of potential suitors by assessing the chemicals emanating from feces. For example, the females prefer males that eat termites (a good source of protein) to those that eat ants (a poor source of nutrients). By determining what a potential mate eats, the female can determine how healthy he is, what a good forager he is, and the quality of the territory he maintains. She thus gets an indication as to whether he will be a good sperm donor for her eggs. It is likely that all *Plethodon,* and perhaps many other of the amphibians within the Great Smokies, rely extensively on their well-developed chemosensory abilities.

Egg brooding is common in many species of salamanders. The function of egg brooding is to deter predators (both of conspecific salamanders, other species of salamanders, and other types of small predators), to ensure that fungus does not envelop the clutch (their skin secretions may contain antibiotic properties), and to remove infertile eggs that might go bad and thereby contaminate the rest of the clutch. Clutches that are not brooded have virtually no chance of survival.

As with other *Plethodon,* this species has a noxious skin secretion that may make it distasteful to small predators. The red stripe of most specimens could be a warning coloration, but it could also assist in camouflage, especially to predators that do not see color. This species is most active when colorful leaves cover the ground.

Southern Zigzag Salamander
Plethodon ventralis

Etymology

Plethodon: from the Greek words *plethore,* meaning fullness or full of, and *odon,* meaning teeth; literally full of teeth. The name reflects the large number of teeth on both the jaws and the roof of the mouth (the vomerine teeth); *ventralis:* a Latin name referring to the uniformly colored pattern on the belly (ventral region) of the species.

Identification

Adults. This is a small dark-colored salamander with a reddish dorsal zigzag stripe down its back. In the Great Smokies, the dorsal zigzag stripe is usually very dark red and may be difficult to see on some animals. There may be some red where the limbs join the body and on the belly. The sides and back other than the stripe are dark with tiny white spots, giving a slight speckled appearance, and the bellies are gray to black. In other regions, a stripeless color pattern is occasionally observed, but all Southern Zigzag Salamanders that I have found in the Great Smokies have at least a faintly zigzag stripe. Costal grooves: 18. Adults measure 6.5–11 cm (2.5–4.3 in) TL.

Larvae. None. This species has direct development.

Southern Zigzag Salamander adult, Whiteoak Sink.

Eggs. Females deposit 2–5 eggs in a protected area, presumably well underground. Only a very few nests of this species have been discovered, and none from the Great Smokies. The eggs are 4–4.5 mm in diameter.

Similar Species. This species most closely resembles the Southern Red-backed Salamander, *P. serratus,* in the Great Smoky Mountains. Southern Red-backs have a brighter red, straight-edged dorsal stripe, and a salt and pepper-colored grayish belly.

Taxonomic Comments. Until recently, the Southern Zigzag Salamander was considered conspecific with the Zigzag Salamander, *P. dorsalis,* and older literature on the amphibians of the Great Smokies (e.g., King 1939a, 1939b; Huheey and Stupka 1967) refer to it under this name. In an analysis of protein variation in this widespread species, the southern and southeastern populations formerly referred to as *P. dorsalis* were found to vary significantly from their Midwestern counterparts. Based on this molecular evidence, *P. ventralis* was recognized as distinct from the Zigzag Salamander (Highton 1997).

Distribution

The Southern Zigzag Salamander occurs in central and southeastern Kentucky, southwestern Virginia, eastern Tennessee, extreme western North Carolina, northeastern Georgia, northern Alabama, and northeastern Mississippi. In the Great Smoky Mountains, this species is only found in

Whiteoak Sink and in the deep sinkhole in the vicinity of Bull Cave. Contrary to earlier reports (Huheey and Stupka 1967), the species does not appear to occur at The Sinks. Repeated efforts to locate it near the park headquarters also have been unsuccessful. However, the species is likely common in the limestone and dolomite valleys immediately to the north and northwest of the park.

Life History

The life history of this species in the Great Smoky Mountains is probably very similar to that of the Southern Red-backed Salamander, which it closely resembles. It is found especially under rocks and, to a lesser extent, surface logs and litter. Eggs are deposited in terrestrial sites, and the young hatch in autumn as miniature adults. Southern Zigzag Salamanders are usually found in the spring (late March to early May) and in the autumn, and disappear during the heat of the summer. Like other woodland salamanders, they eat a wide variety of very small invertebrates.

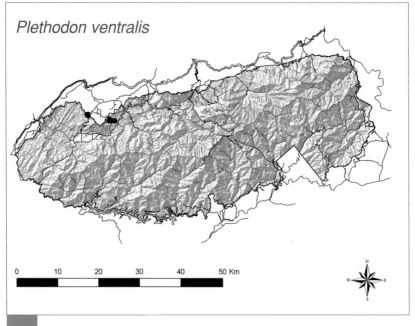

Plethodon ventralis

Distribution of the Southern Zigzag Salamander *(Plethodon ventralis)*.

known specimens of Mud Salamanders from the Great Smokies have been found at this location. There does not appear to be any suitable nearby habitat from which this site could have been colonized.

Remarks

As with Red Salamanders, the coloration of this species may serve as a warning to predators that might confuse the Mud Salamander with the toxic Red-spotted Newt.

Black-chinned Red Salamander
Pseudotriton ruber

Black-chinned Red Salamander Adult, The Sinks.

Etymology

Pseudotriton: from the Greek *pseudes,* meaning false, and *triton,* which refers to the mythological sea god Triton, the son of Poseidon and Amphitrite. The name *triton* also refers to a European genus of newts *(Triturus),* so in effect the common name means "false newt"; *ruber:* Latin, meaning red.

Identification

Adults. This is a stout, bright red salamander with profuse, irregular, black spotting on the dorsum and sides. In older individuals, the spots tend to fuse and the animal may become purplish or melanistic. The

Black-chinned Red Salamander Larva, Little Brier Creek (July 11, 2000).

belly is red to rose and normally unpigmented. Tails are short in relation to the body length, and the eyes are yellow to gold. Juveniles are bright red, but otherwise resemble the adults. Adults measure 9.5–18 cm (3.7–7.1 in) TL.

Larvae. Larvae are streamlined for a stream existence and are light brown above and whitish below. The belly lacks pigmentation. As they grow, the larvae develop a black spotted pattern along the sides. The tiny spots do not form a streaked appearance. As the larvae get older, they become darker. Adult color patterns are attained within a few months of metamorphosis.

Eggs. Females deposit between 29 and 130 eggs on the undersides of rocks and forest litter in springs and seeps. Eggs measure 4 mm in diameter, and are attached singly to the overhanging substrate by a gelatinous stalk. Females may brood their eggs, and more than one female may deposit her eggs at any particular location.

Black-chinned Red Salamander
(Pseudotriton ruber).

Similar Species. This species most closely resembles the Mud Salamander (*P. montanus*) in the Great Smokies. Red Salamanders are usually much more red than orange, have a more diverse spot pattern with larger and more dense concentrations of dorsal spots, and are more squat than the Mud Salamander. Red Salamanders also wander much farther from their breeding streams than do Mud Salamanders; they are therefore more likely to be encountered terrestrially. The larvae of these species are very similar in appearance in small size classes; the Mud Salamander larva usually is more flecked than spotted laterally. Larval Spring Salamanders are uniformly lavender-colored and lack any spots or flecks.

Taxonomic Comments. Of the four subspecies of Red Salamanders (*ruber, schenki, vioscai, nitidus*), the Black-chinned Red Salamander (*P. r. schenki*) is the subspecies that inhabits the Great Smoky Mountains.

Distribution

Red Salamanders occur from southeastern New York to the Gulf Coast, and west to eastern Louisiana, Mississippi, and western Tennessee. They generally avoid the coastal plain, except along the northern Gulf of Mexico, and they are most commonly associated with the uplands and mountains of the Appalachians, Alleghenies, and Cumberland Plateau. The Black-chinned Red Salamander is found in the Great Smokies and other

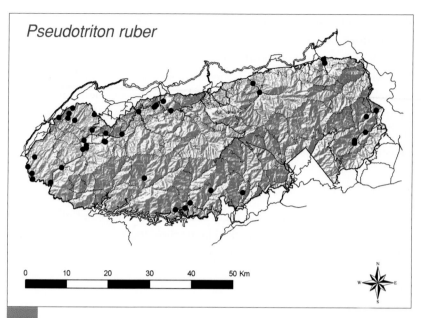

Pseudotriton ruber

Distribution of the Black-chinned Red Salamander *(Pseudotriton ruber)*.

high mountains south of the French Broad River to the intersection of North Carolina, South Carolina, Tennessee, and Georgia. In the Great Smokies, Red Salamanders occur at lower elevations throughout the park, from mesic deciduous forests to the dry oak-pine forests of the west (e.g., the Cane Creek and Beard Cane Creek drainages). During field surveys from 1998 to 2001, Black-chinned Red Salamanders were found from 360 m (1,180 ft) in the Cane Creek drainage to 960 m (3,150 ft) southwest of the Little Cataloochee Church in Little Cataloochee. Huheey and Stupka (1967) stated that they are occasionally found as high as 1,525 m (5,000 ft).

Life History

Red Salamanders breed in seeps and small streams where the larvae will spend from 1.5 to 3.5 years before transforming. Egg deposition occurs in the autumn throughout the winter. In the Smokies, gravid females have been recorded in early September in Cades Cove. Because of their long larval period, larvae may be encountered throughout the year. They live in the leaf litter and other bottom debris and sometimes are found walking on the quiet stream bottoms. Larvae transform throughout the warmer months of the year, but mostly in May and July in western North Carolina. Metamorphosis occurs at 62 to 86 (2.4 to 3.4 in) mm TL. Maturity is reached in about 4 years for males, and 5 years for females. Adults spend most of their life terrestrially, but little is known of their life history away from the breeding streams. They may be encountered at considerable distances from the nearest flowing water.

Abundance and Status

The Black-chinned Red Salamander is a common inhabitant of the Great Smoky Mountains National Park. The streams and forests of the national park provide ideal habitat for this species.

Remarks

The bright red coloration of these salamanders makes them extremely conspicuous. It is now known that they mimic the red eft stage of the Eastern Red-spotted Newt and even have a postural tail display similar to that of the newt (Howard and Brodie 1973). The combination of similar size, coloration, and posture to the highly toxic newt allows them some degree of protection from small predators, especially birds. I have encountered Red Salamanders moving through the leaf litter during the day, usually at times of high humidity or rain, when they are quite visible. If a would-be predator has ever encountered a newt, it is likely to avoid Black-chinned Red Salamanders. Unfortunately, some predators will eat nearly anything. The non-indigenous feral pig, so common to much of the Smokies, does a tremendous amount of damage to the surface litter and probably eats any salamander it finds. My wife and I discovered one Black-chinned Red Salamander near the church in Little Cataloochee on 8 May 1999 that had been killed, presumably by the pigs that had recently been observed in the area.

Frogs

Northern Cricket Frog
Acris crepitans

Etymology

Acris: from the Greek *akris*, meaning locust, a reference to the call of this species, which sounds like an insect; *crepitans:* from the Latin *crepitans*, meaning clattering, again referring to the clicking call of this small frog.

Northern Cricket Frog adult, green phase. Buncombe County, North Carolina. This species also has brownish-tan and red phases. Photo by Richard Bartlett.

Identification

Adults. Northern cricket frogs are small frogs, ca. 19 to 38 mm (0.75 to 1.5 in) in length. Their ground color is gray to tan or brown, and the skin is somewhat bumpy. Dorsally, the species has a dark triangle between the eyes pointing posteriorly, followed sometimes by a median stripe and a Y figure down the back. The dorsum is quite variable, and can have a patch of green, red, yellow, or deep brown to gray in the Y. The rear feet are fully webbed. Adults measure 16–35 mm (0.6–1.4 in) TL.

Larvae. The tadpoles are medium-sized, light to dark gray, and have long tails with low tail fins. They are distinguished by their black tail tips, which no other tadpole in the Great Smoky Mountains region possesses. Large tadpoles range between 30 and 46 mm (1.2 to 1.8 in) in length.

Eggs. The eggs are deposited in small, floating packets and are spread in a thin film on the water's surface, or they may be placed among the vegetation or on the bottom of the water. Although a female may deposit up

Northern Cricket Frog *(Acris crepitans).*

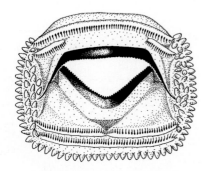

Northern Cricket Frog *(Acris crepitans).*

to 250 eggs in a breeding season, the eggs are deposited at about 1 to 6 at a time. These are 2.3 mm in diameter.

Similar Species. Northern Cricket Frogs are most often confused with Spring Peepers and Upland Chorus Frogs. Spring Peepers and Upland Chorus Frogs are both larger than the Northern Cricket Frog. Spring Peepers are uniformly light to dark brown with a distinctive **X** on the back and no bright coloration. Upland Chorus Frogs are dark brown with three parallel dark dorsal stripes down their back. The Northern Cricket Frog has neither the **X** nor the parallel stripes, and it has a somewhat more "warty" appearance than these smooth species.

Taxonomic Comments. There are three subspecies of *A. crepitans: blanchardi, crepitans,* and *paludicola.* The Northern Cricket Frog, *A. c. crepitans,* is the subspecies found near the Great Smoky Mountains.

Distribution

Northern Cricket Frogs occur from southern New York to the northern Gulf Coast, and west to eastern Texas. They are replaced in the Midwest and Great Plains by Blanchard's Cricket Frog, *A. c. blanchardi,* and in southwestern Louisiana and adjacent southeastern Texas by the Coastal Cricket Frog, *A. c. paludicola.* There is some question as to whether the Northern Cricket Frog has ever occurred within the boundary of the Great Smoky Mountains National Park. King (1944) collected 4 specimens from 266 m (875 ft) at Chilhowee on 14 June 1940 (labeled 13 June in the park's research collection), and Huheey and Stupka (1967) mentioned hearing its call at Laurel Lake (now mostly dry) near Townsend in early April and May. The location of the former town of Chilhowee is now under

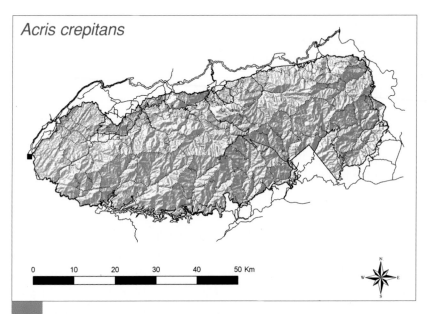

Acris crepitans

0 10 20 30 40 50 Km

Historic distribution of the Northern Cricket Frog *(Acris crepitans)*. The square shows the location of the former town of Chilhowee.

Lake Chilhowee, and repeated call surveys along the northern inlets of the lake have failed to detect this species. It is likely extirpated, even if it did peripherally occur within the park. I include *A. crepitans* in the park's species list only because it doubtless occurs at low elevations (335 m [1,100 ft]) very near the northwest side of the park and in Tuckaleechee Cove.

Life History

Nothing is known about the life history of this species in the vicinity of the park. Huheey and Stupka (1967) mentioned that it called in April and May in Tuckaleechee Cove near Townsend. In other parts of its range, it is known to emerge from hibernation well before the breeding season. The mating call is a loud "crick-crick" repeated over and over. Egg laying occurs from April throughout the summer, and hatching occurs in a few days. The tadpoles take ca. 35 to 70 days to metamorphose, usually from June until October depending on when the eggs were deposited. Northern Cricket Frogs are usually abundant along favorable shorelines and do not stray far from water. They are expert hoppers, and make long, rapid jumps despite their small size. They eat a wide variety of small invertebrates that are found in the grassy margins of shallow ponds, lakes, small reservoirs, and similar habitats, especially those with grassy cover.

Abundance and Status

As stated above, this species probably does not occur within the park. It is a common species throughout much of its range, although cricket frogs in the upper Midwest have declined dramatically since the early 1980s, even though the habitat has remained largely intact.

Remarks

Northern Cricket Frogs usually call from small ponds in open fields, such as wet pastures, and from around lake margins and in open-canopy swampy situations. This species might have occurred in the creek bottoms of the northwestern side of the park (e.g., Cane Creek, Beard Cane Creek) when these areas were open fields. However, as succession has taken place, suitable habitat no longer exists. There are no apparent reasons why this species could not survive in Cades Cove, although extensive call surveys through the years have failed to locate it there.

American Toad adult, Cades Cove.

Eastern American Toad
Bufo americanus

Etymology

Bufo: Latin for toad; *americanus:* Latinized name for America, meaning of or belonging to America.

Identification

Adults. This is a medium to large toad; females may be much larger than males. The dorsum is gray or a rich brown to nearly brick red, and is covered with knobs and bumps; the parotoids (bean-shaped granular glands behind the eyes) and cranial crests (bony ridges between and just behind the eyes) are prominent; and there is usually one knob or wart per dark spot. The bellies are white with some mottling around the sides. Males have dark throat pouches. Adults measure ca. 50 to 110 mm (2 to 4.3 in) TL.

American Toad (*Bufo americanus*).

Larvae. Tadpoles are small and uniformly colored dark brown to black. The belly may have an aggregate of gold or copper spots. Tails are short and bicolored.

Eggs. The eggs are deposited in two long strings, one arising from each oviduct. Egg strings are placed on the pond bottom, or they may be entangled in the vegetation. These strings can be very long (60 m [197 ft]) and contain 4,000–12,000 eggs.

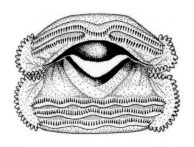

American Toad (*Bufo americanus*).

Similar Species. Toad tadpoles are difficult to tell apart, especially when they are small. Although there may be some overlap, early season toad tadpoles are most likely *B. americanus,* whereas those found later in the season are *B. fowleri.* Adults also may be difficult to tell apart, but *B. fowleri* has much smaller cranial crests, less-prominent parotoids, a mid-dorsal white stripe, and 2 or 3 or more knobs or warts per dark dorsal spot.

American Toad egg strings, Cades Cove.

Adult females are also much smaller than *B. americanus* females.

Taxonomic Comments. There are two recognized subspecies of American Toads, *Bufo americanus americanus* and *B. a. charlesmithi.* The Eastern American Toad *(B. a. americanus)* occurs in the Great Smoky Mountains. The American Toad hybridizes with many other species of North American toads; thus, there may be several evolutionary lineages within the taxon.

Distribution

The Eastern American Toad occurs from southern Labrador to the mountains of northern Georgia and Alabama, and from the Midwest to the eastern Great Plains north to eastern Manitoba. It is replaced in the Ozark Highlands, western Kentucky and Tennessee, and southern Indiana and Illinois by the Dwarf American Toad, *B. a. charlesmithi*. In the Great Smoky Mountains, the Eastern American Toad breeds in a wide variety of mostly lowland stream shallows, woodland pools, ponds, and other wetlands, particularly in Cades Cove, where breeding sites are common. However, individuals may wander extensively to higher elevations throughout the park during the non-breeding season. Huheey and Stupka (1967) mention breeding occurring at 1,524 m (5,000 ft) in the Flat Creek area and noted 16 observations above this elevation. They even recorded it "within one-half mile of the summit Clingmans Dome." Most of the sightings recorded during surveys from 1998 to 2001 were lower than 914 m (3,000 ft), although one toad was observed on Welch Ridge at Bear Creek at 1,492 m (4,894 ft) on 8 August 2000. Eastern American Toads also frequent cave entrances during the summer (e.g., Dodd, Griffey, and Corser

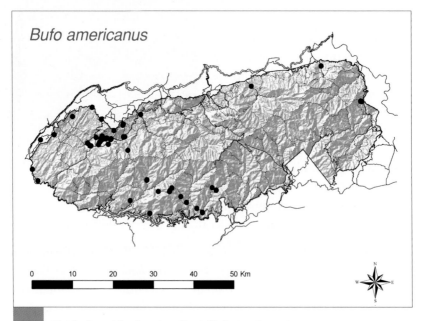

Bufo americanus

Distribution of the American Toad *(Bufo americanus)*.

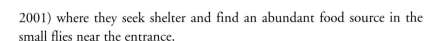

2001) where they seek shelter and find an abundant food source in the small flies near the entrance.

Life History

American Toads breed in the spring, primarily in March and April. Adults move to breeding sites, which might include any place from a small road-side ditch to a stream or a pond, where males set up loud trilling choruses to attract females. Females only remain at a breeding site long enough to mate and deposit their eggs. Eggs are deposited as two long strings that hatch within 3 to 12 days into tiny gilled larvae. The gills quickly become internalized, and the larvae then are recognizable as tadpoles. The earliest we have seen eggs and tadpoles of this species in the Smokies is 8 March, although calling males have been heard as early as 14 February and the earliest record for breeding within the park is 17–27 February (Huheey and Stupka 1967). Tiny metamorphs (7–12 mm [0.3–0.5 in] TL) appear from early July to mid-August after 50 to 65 days as tadpoles. They grow rapidly, however, and are much larger juveniles by September.

Individual American Toads probably are active in nearly all months of the year in the Smokies, depending on temperature. King (1939b) noted emergence in December and January. Jeff Corser (pers. comm.) has seen them by early February (11 February 1999) along the Hannah Mountain Trail above Cades Cove. According to Huheey and Stupka (1967), emergence in great numbers normally occurs in March. These authors stated that the latest date American Toads had been recorded breeding was 18 April at 1,524 m (5,000 ft) along Flat Creek, but we have heard adults calling on 7 May (1999), with egg deposition occurring on 9 May, at Gourley Pond in Cades Cove. Large numbers of individuals may form a single chorus: on 7 April 2001, Jeff Corser heard hundreds of males in full chorus, during warm and sunny conditions, at the mouth of Hazel Creek where it enters Fontana Reservoir.

Abundance and Status

Of the two toads in the park, American Toads are much more widespread and abundant. During surveys from 1998 to 2001, we commonly observed American Toads along trails and in the forest far from the nearest possible breeding site.

Remarks

Like all toads, American Toads have noxious and toxic skin secretions located in their dorsal skin glands, particularly in the large parotoid glands located behind the head and in the knobs on the back. The purpose of these secretions is to deter predators. However, raccoons sometimes learn to avoid the dorsal skin surface by attacking the toad through its belly; when this occurs, large numbers of toads may be killed at a breeding site. Toads are gentle and harmless to people, and they display considerable variation in personality. However, small children and pets should avoid handling these species. Persons should always wash their hands after handling a toad, not to prevent warts (the belief that contact with a toad's skin causes warts is a myth), but to prevent the secretion from irritating the eyes.

During courtship, male toads grab nearly any appropriately sized object at a breeding pond and firmly hold on. The grasp in which a male holds onto a female during courtship is termed amplexus, and it allows the male to fertilize the female's eggs as she deposits them. All frogs in the Great Smokies have a similar behavior and, as such, fertilization is external. When picked up and held behind the front legs (simulating amplexus), males emit a "warning chuckle" consisting of a series of squeaks and rapid vibrations. This behavior alerts a courting male that the toad so grabbed is actually a male rather than a female. It means "Let go!" In this way, males at breeding ponds avoid amplexus by other males. If a toad does not squeak when held, a courting male will know that he is courting a female. Trouble ensues when males try to court other species or inanimate objects that don't emit a warning chuckle. There are anecdotes in the scientific literature of male toads amplexing apples, rocks, and old shoes, as well as other species, such as leopard frogs.

Fowler's Toad
Bufo fowleri

Etymology

Bufo: Latin for toad; *fowleri:* Latinized name in honor of New Jersey mineralogist Samuel Page Fowler (1800–1888).

Identification

Adults. This is a medium-sized toad. The dorsum is gray to brown, and it is covered with knobs or warts and bumps; the cranial crests (bony ridges between and just behind the eyes) are not very prominent; the parotoids (bean-shaped granular glands behind the eyes) are obvious but not as large as those in *B. americanus;* there is usually more than 2 or 3 knobs or warts per dark spot on its dorsum. Many individuals have a light line down the middle of the back. The bellies are whitish and unspotted. Males have dark throat pouches. Adults measure ca. 45 to 82 mm (1.7 to 3.2 in) TL.

Fowler's Toad adult, Cataloochee Creek at gaging station.

Fowler's Toad (*Bufo fowleri*).

Larvae. Tadpoles are small and uniformly colored dark brown to black. Tails are short and *not* distinctly bicolored.

Eggs. The eggs are deposited in two long gelatinous strings, one arising from each oviduct. Egg strings are entangled in the vegetation. These strings contain between 5,000 and

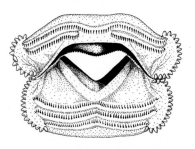

Fowler's Toad (*Bufo fowleri*).

10,000 eggs, with about 17 to 25 eggs per 25 mm (1 in). They are more compact than the egg strings of *B. americanus* and are rarely more than 3 m (ca. 10 ft) long.

Similar Species. Toad tadpoles are difficult to tell apart, especially when they are small. Although there may be some overlap, early season toad tadpoles are most likely *B. americanus,* whereas those found later in the season are *B. fowleri.* Adults of the two species also may be difficult to tell apart, but *B. americanus* has much larger cranial crests, prominent parotoids, and usually only a single knob or wart per dark dorsal spot. Adult *B. americanus* females are also much larger than *B. fowleri* females.

Taxonomic Comments. This species has often been referred to as a subspecies of Woodhouse's Toad, *Bufo woodhousii* (i.e., as *B. w. fowleri*), because it hybridizes with this toad in eastern Texas and Oklahoma.

Distribution

Fowler's Toad occurs from New Hampshire, Massachusetts, and the Great Lakes region south to the Gulf Coast, and west to eastern Texas, Oklahoma, and Missouri. In the southwestern part of its range, extensive areas of hybridization occur with Woodhouse's Toad. In the Great Smoky Mountains, Fowler's Toad occurs primarily in the lowlands. During field surveys from 1998 to 2001, it was recorded from Cades Cove, Big Cove, Big Spring Cove, along Cataloochee Creek (near the gaging station), lower Noland Creek, lower Eagle Creek (Fontana Lake to Horseshoe Bend), and Clark Branch. It likely occurs at the lower ends of all creeks flowing into the north shore of Fontana Lake. The highest elevation at which Fowler's Toad was found during these recent surveys was 975 m (3,200 ft) along Gunna

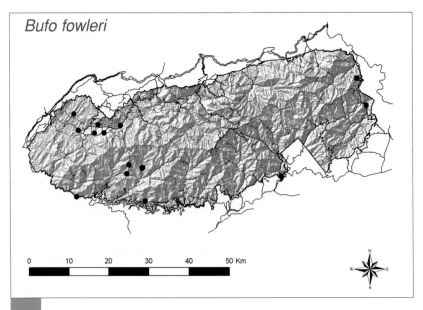

Distribution of Fowler's Toad *(Bufo fowleri)*. The square shows the historic location on Mount Sterling Ridge.

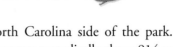

Creek (Eagle Creek drainage) on the North Carolina side of the park. According to Huheey and Stupka (1967), it occurs sporadically above 914 m (3,000 ft), with the highest elevation ever recorded at Mt. Sterling Ridge at 1,219 m (4,000 ft) (King 1939b).

Life History

Fowler's Toad breeds somewhat later in the spring, and throughout the summer, in comparison with the Eastern American Toad. This toad also has a loud, trilling call, but it is of much shorter duration than that of the Eastern American Toad. King (1939b) records them breeding at the same ponds at the same time, thus making identification of eggs and small tadpoles difficult. We have recorded egg strands as early as 11 May (2000) in Big Cove on the North Carolina side of the park, and as late as 5 August (1999) along Abrams Creek. King (1939b) recorded the earliest breeding date as 30 March (1938) near Townsend at the park's boundary. The larval period lasts 40–60 days. Tiny tadpoles were even found 30 August 2000 in a stump hole along Beard Cane Creek, indicating an almost fall breeding season for this species. Nearly all the metamorphs (7.5–11.5 mm [0.3–0.5 in] TL) that we observed in field surveys were seen in early to mid-August; small juveniles are commonly seen in September.

Fowler's Toads wander extensively afield, but they may not travel as much as the Eastern American Toad (Huheey and Stupka 1967). Adults may be seen anytime from at least March into the fall before the first frosts; there does not appear to be any winter activity.

Abundance and Status

Fowler's Toad appears to be less common in the Great Smoky Mountains than the American Toad. It is widely distributed within the park, however, and is not considered to be a rare species.

Remarks

Fowler's Toads have noxious and toxic skin secretions located in their dorsal skin glands, particularly in the large parotoid glands located behind the head and in the knobs on the back. The purpose of these secretions is to deter predators. Although the secretions are biochemically complex, toad secretions contain digitalis-like substances that have been used in the treatment of heart ailments. Even the eggs and tadpoles of some species are toxic. Small children and pets should avoid handling these species. Persons should always wash their hands after handling a toad, not to prevent warts (the belief that contact with a toad's skin causes warts is a myth), but to prevent the secretion from irritating the eyes. Toads are gentle and harmless to people, and both American and Fowler's Toads display considerable variation in personality.

Eastern Narrow-mouthed Toad adult, Abrams Creek Ranger Station.

Eastern Narrow-mouthed Toad
Gastrophryne carolinensis

Etymology

Gastrophryne: from the Greek *gastros,* meaning belly, and *phryne,* meaning toad. The name may refer to the fat little belly of this species; *carolinensis:* Latinized noun referring to Carolina. The species was described from specimens found near Charleston, South Carolina. Although commonly called a toad, this species is in the family Microhylidae, not the family Bufonidae as are the true toads.

Identification

Adults. The Eastern Narrow-mouthed Toad is a small, pointy-snouted frog. The dorsum is brownish red to blue-black, and may be highly mottled with coppery or silvery pigmentation. A reddish (or chestnut) color may give the appearance of dorso-lateral bands; the reddish pigmentation also may be present on the front legs. There is a small fold across the dorsum just behind the eyes. The belly is medium to dark gray, and males have a black throat where the vocal pouch is located. There is no webbing between the toes. Adults measure 22–38 mm (0.9–1.5 in) TL.

Larvae. Tadpoles of this species are small and jet black, with lateral white to pink stripes on either side of the posterior portion of the body extending onto the tail musculature. Viewed from the side, the head comes to a

fairly sharp point. Unlike other tadpoles in the Great Smokies, the jaws do not have keratinized teeth, and no oral disk is present.

Eggs. The eggs are deposited in a small surface film with a mosaic structure; the egg mass is round or squarish, with 10 to 150 eggs deposited in each mass. Breeding occurs in a wide variety of standing-water localities, usually with an open tree canopy, but not in deep water.

Similar Species. This small frog cannot be confused with any other frog in the Great Smoky Mountains. It lacks toe webbing, has no warts, is small, and has a very pointed short

Eastern Narrow-mouthed Toad
(Gastrophryne carolinensis).

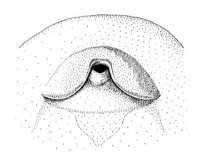

Eastern Narrow-mouthed Toad
(Gastrophryne carolinensis).

snout. Tadpoles most closely resembles toad *(Bufo)* tadpoles, but are more dorso-ventrally flattened and usually have the light colored lateral stripes.

Taxonomic Comments. No subspecies are recognized.

Distribution

The Eastern Narrow-mouthed Toad is found from the Delmarva Peninsula throughout the southeastern United States, and it occurs to the west in eastern Texas and Oklahoma. It is generally absent from the Appalachian Mountains. In the Great Smoky Mountains National Park, this species is known only from the Abrams Creek drainage in Cades Cove (at Shields Pond, a shallow springfed pond in the field northwest of the Tipton Oliver place) and in the vicinity of the Abrams Creek Ranger Station. All locations are below 550 m (1,800 ft). Eastern Narrow-mouthed Toads also are found in some of the valleys and coves directly to the north of the park.

Life History

Males chorus in the summer, and they have been heard from 1 June to 5 September (2000) in Cades Cove. The calls are quite distinctive, sounding

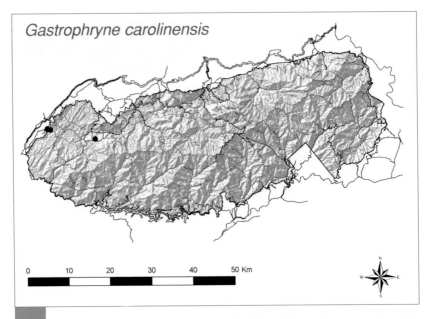

Distribution of the Eastern Narrow-mouthed Toad *(Gastrophryne carolinensis).*

like a high-pitched "baaaa" of a sheep. The larval period is from 20 to 70 days, based on literature records. Very small tadpoles have been seen by 10 July (2000), and the metamorphs measure 8.5–12 mm (0.3–0.5 in) TL. Longevity is ca. 4 years, although few animals likely reach this age. This little narrow-headed frog feeds almost exclusively on ants. They also eat small termites.

Abundance and Status

Nothing is known about the status of this species within the park, about its biology, or about how it uses habitats during the nonbreeding season. Data from studies elsewhere in the Southeast suggest that it probably remains within several hundred meters of the breeding site.

Remarks

As with some other frogs, this species has a noxious skin secretion which may help to deter predators and to make them impervious to the stings of the ants on which they feed.

Cope's Gray Treefrog
Hyla chrysoscelis

Cope's Gray Treefrog adult, road to Cosby campground.

Etymology

Hyla: Greek for Hylas, the companion of Hercules. Hylas was one of the mythological Argonauts who sailed with Jason looking for the Golden Fleece. At a spring, a water maiden pulled Jason in, after which he wandered about calling Hylas' name. Exactly what Laurenti had in mind when he coined the name is unclear; *chrysoscelis:* from the Greek *chrysos,* meaning gold, and *kelis,* meaning spot or stain. The name refers to the golden spots on the back of the thigh of this gray treefrog. The name Cope, in the common name, refers to famed herpetologist and paleontologist Edward Drinker Cope (1840–1897), who described this species in 1880.

Cope's Gray Treefrog larva, Feezell Branch (September 2000).

Identification

Adults. Cope's Gray Treefrog is a small to medium-sized frog with a distinctive lichenlike dorsal pattern. The lichen coloration is composed of various gray to buff patches; sometimes the patches may be distinctively greenish in color. There is a

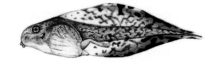

Cope's Gray Treefrog
(Hyla chrysoscelis).

light patch underneath the eye. The toes are tipped by conspicuous toe pads; the rear feet are partially webbed. The inner thigh is bright orange and unspotted. Bellies are unpigmented. Adults measure ca. 32 to 62 mm (1.25 to 2.4 in) TL.

Larvae. Tadpoles are small to medium-sized and gray, with a high dorsal tail fin. The tail is long with black blotches, and the height of the tail fin equals the height of the tail musculature. As tadpoles mature, a bright orange to red background color develops on the tail fin.

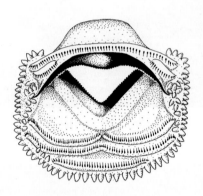

Cope's Gray Treefrog
(Hyla chrysoscelis).

Eggs. Eggs are deposited in a small surface film, but there is no mosaic structure as there is in the Eastern Narrow-mouthed Toad. Females lay 5–40 eggs per mass in shallow ponds, woodland pools, and along the margins of deeper ponds. Eggs may be loosely attached to submerged aquatic vegetation or they may float free.

Similar Species. In the Great Smoky Mountains, this species cannot be confused with any other species. It is the only true treefrog in the mountains and can be readily identified by its greatly expanded toe tips.

Taxonomic Comments. Although this species is extremely similar in appearance to the Gray Treefrog *(Hyla versicolor)* and may have given rise to that species (Ptacek et al. 1994), Cope's Gray Treefrog is well differentiated by its call, karotype, and cell volume. Various evolutionary lineages have been described, but there are no published reports suggesting that the species should be split taxonomically. In areas where the ranges of these species overlap or where identifications are questionable, some authors prefer to recognize the Gray Treefrog complex rather than apply a specific name to an individual frog. Inasmuch as *H. chrysoscelis* occurs throughout eastern Tennessee and most of western North Carolina (Burkett 1989; Redmond and Scott 1996), treefrogs in the Great Smoky Mountains National Park are referred to this species. Burkett's (1989) suggestion that *H. versicolor* might one day be found at high elevations in the Great Smokies remains speculative.

Distribution

The Gray Treefrog complex occurs from southern Canada (Québec to Manitoba) south to northern Florida and the Gulf Coast, and west to east-central Texas and the eastern Great Plains. In the southeastern United States,

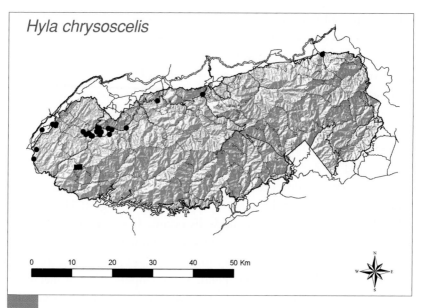

Distribution of Cope's Gray Treefrog *(Hyla chrysoscelis)*. The squares show historic locations at Sheep Pen Gap and Gregory's Bald.

Cope's Gray Treefrogs appear to be much more widespread than the Gray Treefrog. In Great Smoky Mountains National Park, USGS survey crews found *H. chrysoscelis* only at lower elevations on the Tennessee side of the park between 1998 and 2001. Cope's Gray Treefrog most commonly occurs at elevations below ca. 610 m (2,000 ft). Localities at which the species was seen or heard calling include the road into Cosby campground, Sugarlands, The Sinks, along Little River Road near Metcalf Bottoms, Big Spring Cove, Cades Cove (throughout), lower Abrams Creek near the campground, along lower Panther Creek, and along the eastern shore of Chilhowee Lake. According to Huheey and Stupka (1967), however, individuals have been heard as high as Gregory Bald (1,508 m [4,948 ft]), Cataloochee Ranch (1,490 m [4,888 ft]) and Sheep Pen Gap (1,405 m [4,610 ft]). Thus, non-breeding individuals may disperse great distances from breeding ponds.

Life History

Calling males have been heard from 11 April (2000) through 12 July in Cades Cove, although King (1939b) reports calling as early as 31 March.

The mating call is a short, fluted, rapidly produced trill. Males call from the trees as they move to breeding ponds, and remain in the vicinity of the pond, calling from surrounding vegetation. In late summer, they move back to terrestrial sites. Males have a variety of calls, such as warning, mating, and rain calls. These serve to advertise their presence, establish territories, warn rival males, and attract females. Indeed, females probably can assess the "quality" of a male (for example, his size and quality of his calling territory) by his call. Females remain at the ponds long enough to mate and lay eggs, which are deposited in a variety of wetlands, from small temporary puddles to the large sewage pond in Cades Cove. Eggs hatch rapidly, and tadpoles are present from July to August. The tadpoles take 45–65 days before metamorphosis occurs. On 7 August 2000, my wife and I observed hundreds, possibly thousands, of Cope's Gray Treefrog metamorphs (13–20 mm [0.5–0.8 in] TL) departing from the nearly dry pond bed at Gourley Pond. Many tadpoles did not metamorphose in time, and lay rotting in mud puddles. Transforming larvae also were seen at other Cades Cove wetlands on this date. Small juveniles were seen at Gum Swamp on 20 September (1999), indicating rapid growth prior to the onset of their first winter. When handled, this species proves to be quite sticky.

Abundance and Status

Although no quantitative data are available, this species appears to be common in the lower elevations of the park.

Remarks

On 12 July 1999, my wife, Kelly Irwin, and I observed a small DeKay's Brownsnake *(Storeria dekayi)* foraging among recently deposited eggs of *Hyla chrysoscelis* at a very temporary puddle that formed after heavy rains at the Abrams Falls parking lot. We could not determine whether the snake was eating the eggs or hunting the tiny tadpoles that were present. Brownsnakes have been reported to eat frog eggs in other areas.

A late spring to early summer walk up the road to Gregory's Cave in Cades Cove, with the new flowers in bloom and Cope's Gray Treefrogs trilling and resonating from the surrounding treetops, is one of the many little pleasures that results from studying amphibians in the Great Smokies.

Spring Peeper
Pseudacris crucifer

Etymology

Pseudacris: from the Greek *pseudes,* meaning false or deceptive, and *akris,* meaning locust. The name literally means "false Acris" and refers to the cricket frogs, genus *Acris.* A false cricket frog; *crucifer:* from the Latin *crucis,* meaning cross, and *-ifer,* meaning bearer. The scientific name refers to the **X** on the back of this small frog.

Identification

Adults. Adult Spring Peepers are light tan to dark brown with a distinctive dark-colored **X** on the back. In a few individuals, the arms of the **X** may approach one another but not actually come into contact. Note that the frogs may appear darker during the day than at night. A dark line connects the eyes. The belly is light and normally unmarked, although a few animals may have a slightly spotted venter. Toe tips are only faintly expanded, and there is no webbing between the toes. Males have a black vocal pouch and are smaller than females. Adults measure 19–35 mm (0.7–1.4 in) TL.

Spring Peeper adult, Cades Cove.

Spring Peeper adults, amplexing pair. Big Spring Cove.

Spring Peeper larva, Big Cove (July 9, 2000).

Larvae. Tadpoles of this species are small and deep-bodied with a medium-sized tail. The tail musculature is mottled, but the fins are clear or with blotches. There are no dots on the

Spring Peeper (*Pseudacris crucifer*).

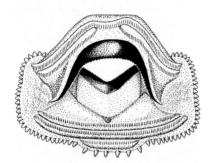

Spring Peeper (*Pseudacris crucifer*).

grayish to light brown body. When viewed from above, the snout has a squarelike appearance.

Eggs. The eggs are deposited singly among the vegetation in shallow water near the bottom. There may be up to 900 eggs laid by any individual female during the breeding season.

Similar Species. Spring Peepers are most likely to be confused with the similarly-sized Upland Chorus Frog. Upland Chorus Frogs have 3 parallel dark bands down their backs rather than an **X**; they are darker in color and the cream to white belly usually is in stark contrast to the darker dorsal coloration; the chorus frog has a distinct white line on the upper lip.

Taxonomic Comments. There are two recognized subspecies of Spring Peeper; the Northern Spring Peeper, *P. c. crucifer,* occurs in the Great Smoky Mountains. Older literature lists this species within the genus *Hyla* rather than *Pseudacris.*

Distribution

Spring Peepers occur from central Canada (Manitoba to Québec and the Maritimes) south to the Gulf Coast and westward to the eastern margin of the Great Plains. In the Great Smokies, they are generally found at lower elevations in the Little River, Abrams Creek, Cane Creek, Beard Cane Creek, and West Prong of the Little Pigeon River drainages. They occur in Cades Cove, Big Cove, Greenbrier, Cataloochee, and Bone Valley. Spring Peepers probably occur nearly anywhere a small woodland pool or wetland is available for breeding. Adults may wander considerable distances from breeding ponds. Huheey and Stupka (1967) stated that it occurs "to the summits of the highest mountains," but noted that it was scarce above 1,220 m (4,000 ft).

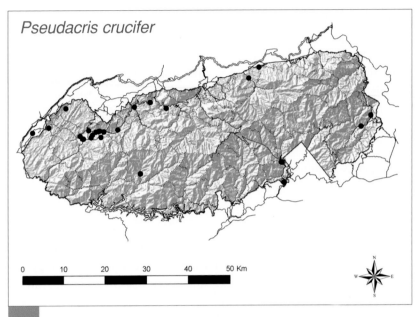

Pseudacris crucifer

| 0 | 10 | 20 | 30 | 40 | 50 Km |

Distribution of the Spring Peeper *(Pseudacris crucifer)*.

Life History

The familiar high-pitched "peep-peep" calls of this species are heard most frequently from late winter (e.g., 11 February 1998) through July, although singular calls are heard virtually throughout the year if cool, rainy conditions prevail. On cold, rainy spring nights, choruses of Spring Peepers occur in great numbers throughout Cades Cove. A few animals are likely to call anytime during the summer and into the autumn; my wife and I have heard them at Gum Swamp in late September during rains. Attempts to locate the frogs in their arboreal habitats have proved difficult, as they are not amenable to arboreal trapping, such as by using PVC pipes. After deposition, the eggs hatch in a few days and tadpoles metamorphose in 3 to 4 months. Tadpoles appear from early April to June in the Great Smokies, and USGS field crews have observed large numbers transforming on 11 May (2000) at Gourley Pond, 5 June (2000) in Gum Swamp, 9 July (2000) in Big Cove, and 8 August (2000) in Bone Valley. The young frogs measure 9–14 mm (0.35–0.5 in) TL at metamorphosis. In the summer and autumn, males sometimes call from high in the trees, but during the breeding season they call from grassy clumps around the margins of woodland pools, grassy swales, and ponds. They are very difficult to locate despite their loud call. Spring peepers are only rarely encountered outside of the breeding season in their terrestrial habitats. Huheey and Stupka (1967) recorded predation on this species by a spotted skunk *(Spilogale putorius)* in November at the Sugarlands Pond behind the National Park Service Visitor's Center.

Abundance and Status

This species is quite abundant in Cades Cove and elsewhere in the park and, indeed, throughout the eastern United States.

Remarks

The Spring Peeper is familiar to millions of people in eastern North America as a harbinger of spring. Like many frogs, it has different types of calls depending on the situation, including mating calls (to attract females) and aggressive calls (to warn away rival males). It is unknown why they call from trees after or preceding the breeding season, although the calls may define the home range. Subtle differences make the calls distinct. For example, a "peep, peep" may trail off into a short trill; this could indicate that a rival is too near and may alert him to give more space to the principal caller in the immediate area. Spring Peepers also are known to call in twosomes (duets) and threesomes (triplets). The first frog calls, followed immediately by the second and third, always in the same sequence.

Upland Chorus Frog adult, Cades Cove.

Upland Chorus Frog
Pseudacris feriarum

Etymology

Pseudacris: from the Greek *pseudes,* meaning false or deceptive, and *akris,* meaning locust. The name literally means "false Acris" and refers to the cricket frogs, genus *Acris.* A false cricket frog; *feriarum:* from the Latin *feriarum,* meaning holidays or leisure.

Identification

Adults. This is a small, ground-dwelling frog that is medium to dark brown or gray with three very dark parallel bands down its back. The upper lip has a white line on it, there may be a dark band extending through the eye and continuing just above the forelimb, and the belly is white to cream color in obvious contrast to the darker dorsal color. Sometimes there may be some dark coloration on the chest. The toes have almost imperceptible expanded pads, and webbing between toes is virtually lacking. Adults measure 19–35 mm (0.75–1.4 in) TL.

Upland Chorus Frog
(Pseudacris feriarum).

Upland Chorus Frog
(Pseudacris feriarum).

Larvae. The tadpoles are small and olive to nearly black in coloration, and they have a bronze belly. The tail is neither long nor short, but medium in length.

Eggs. Eggs occur in a loose, irregular cluster (ca. 60 per cluster) deposited around submerged vegetation. During the course of a breeding season, an individual female may oviposit 1,000 eggs. Upland Chorus Frogs prefers shallow, open-canopied, grassy pools in which to lay their eggs.

Upland Chorus Frog eggs, Cades Cove.

Similar Species. Spring Peepers are most likely to be confused with the similarly sized Upland Chorus Frog. Upland Chorus Frogs have 3 parallel dark bands down their backs rather than an X; they are darker in color, and the cream to white belly usually is in stark contrast to the darker dorsal coloration; and the chorus frog has a distinct white line on the upper lip.

Taxonomic Comments. Some authors consider the Upland Chorus Frog, *P. feriarum,* to be a subspecies of the chorus frog *P. triseriata* (i.e., *P. t. feriarum*). Hedges (1986) suggested that this subspecies merited specific status based on an analysis of 33 genetic loci within 30 northern Temperate Zone hylid frogs. Additional molecular analysis, especially in contact zones between species, may help to clarify chorus frog systematics (Crother et al. 2000). Following Hedges (1986), there are two subspecies of *P. feriarum;* the Upland Chorus Frog, *P. f. feriarum,* is the chorus frog in the Great Smoky Mountains.

Distribution

The Upland Chorus Frog occurs from northern New Jersey/southeastern Pennsylvania south to eastern Texas, southeastern Oklahoma, and Arkansas. It usually occurs in the Piedmont above the Fall Line, although there are exceptions. It occurs throughout the Mid-South, including all of Tennessee and most of Kentucky. Upland Chorus Frogs apparently are absent from western North Carolina. In the Great Smoky Mountains, Upland Chorus Frogs are found abundantly throughout Cades Cove. Necker (1934) also

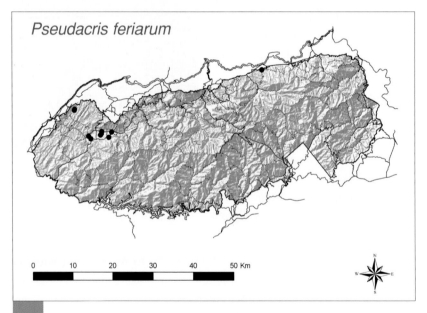

Distribution of the Upland Chorus Frog *(Pseudacris feriarum).*

records it from Greenbrier at 488 to 518 m (1,600 to 1,700 ft), Jeff Corser found it just north of the park boundary at Soak Ash Creek east of Greenbrier, and King (1939b) found it as high as 762 m (2,500 ft) at Fighting Creek Gap. There are two specimens in the park's collection taken at Elkmont on 20 January 1937, but we could not locate suitable habitat for this species in our recent surveys of this area. In the 1930s, Elkmont was far more open than it is today, now that the forest has reclaimed former fields, pastures, and areas of human-maintained clearings.

Life History

This is among the first frogs to call in the winter and into the spring. Breeding stops rather quickly as soon as the weather begins to warm up. For example, large numbers of egg masses and calling males were observed in Cades Cove on 26 January 1999; the latest that calling was recorded from 1998 to 2001 was 7 May (1999). Males call from and females deposit eggs in shallow, grassy wetlands, which are common especially in Cades Cove. According to King (1939b), males call both in the early morning and from the late afternoon into the evening. The calls are easily identified; they sound as if a person is running their fingers along the teeth of a plastic comb.

Egg clutches also have been observed in deep pools located within the grassy fields of Cades Cove (e.g., on 8 April 2000). Tadpoles are present from mid-April to early June. Tadpoles develop rapidly, and the larval period lasts 50–60 days. We have observed recent metamorphs (8–12 mm [0.3–0.5 in] TL) in early June (5 June 2000) at Gum Swamp in Cades Cove. Upland Chorus Frogs are only rarely encountered outside the breeding season. We occasionally have seen individuals under surface debris in drying wetlands, for example, on 8 July, 7 September, and 21 September in Gum Swamp. Even at this time, occasional solitary calls are sometimes heard.

Abundance and Status

In Cades Cove, this species appears to be particularly common. At some historic locations, however, such as Elkmont, the Upland Chorus Frog apparently has disappeared as vegetational succession occurred in conjunction with forest recovery. As long as the open character of Cades Cove is maintained, Upland Chorus Frogs should survive in the Great Smokies.

Remarks

Although easy to hear, these small frogs are almost impossible to locate as they call from grass clumps within their shallow pools. Even when you are absolutely certain you know exactly where the male is, he is usually not where you think he is. Loud choruses are heard in Cades Cove in March and April, and it is rather wonderful to listen to the deafening choruses of Spring Peepers, Upland Chorus Frogs, and Eastern American Toads on a cold, misty, moonlit night.

American Bullfrog Adult, Great Smoky Mountains National Park (exact location unknown). Photo by Richard Bartlett.

American Bullfrog Larva, Abrams Creek near Ranger Station.

American Bullfrog
Rana catesbeiana

Etymology

Rana: Latin for frog; *catesbeiana:* named for Mark Catesby (1679/83–1749), an English naturalist who described the fauna and flora of the southeastern coastal plain and the Bahamas in his book *Natural History of Carolina, Florida & the Bahama Islands,* issued in parts from 1729 to 1747.

Identification

Adults. American Bullfrogs are large, light olive, dark green, or brown, heavy-bodied frogs that have a large tympanum and lack a dorso-lateral fold. They may be distinctly mottled or nearly uniform in coloration dorsally. Distinct black bars are present on the dorsal surface of the thighs and lower legs. The bellies are cream-colored to white with dark reticulations or mottling. The eyes are large, and the rear feet are highly webbed, although the front feet are not. Adults measure 85–200 mm (3.3–8 in) TL. This is the largest frog in the Great Smoky Mountains.

American Bullfrog *(Rana catesbeiana)*.

Larvae. Tadpoles of this species are large, and grayish-green to olive in coloration. There are small widely spaced black flecks throughout the body and tail. The belly is off-white or cream-colored.

Eggs. Eggs are laid in a large surface film in the form of a disk, with 10,000 to 12,000 eggs per disk. The film is placed within vegetation or among sticks and branches.

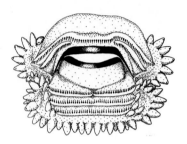

American Bullfrog *(Rana catesbeiana)*.

Similar Species. A combination of large size and a lack of any dorso-lateral fold distinguish this species from any other large green frog (Northern Leopard, Pickerel, Northern Green) in the Great Smoky Mountains. The tadpoles of these four species sometimes are difficult to differentiate from one another (see the species accounts for descriptions of the tadpoles).

Taxonomic Comments. No subspecies are currently described, although there is considerable variation in tadpoles and calls of American Bullfrogs from different geographic regions of North America.

Distribution

American Bullfrogs occur naturally throughout eastern North America, from southern Canada west to the central Great Plains. They have been introduced nearly everywhere in western North America—and indeed in many parts of the world—where they pose a considerable threat to native amphibians and other animals. American Bullfrogs are found in many lowland areas of the park, especially in Cades Cove, in low-elevation streams (Abrams Creek, Cane Creek, Beard Cane Creek, Little Pigeon River, Little River), in small woodland pools (at Sugarlands), along the eastern shore of

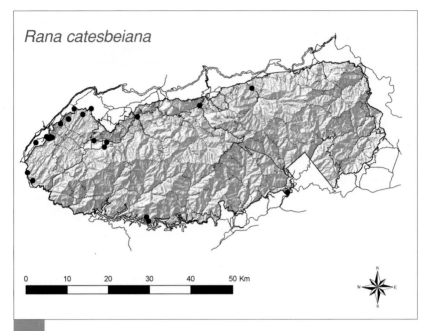

Distribution of the American Bullfrog *(Rana catesbeiana)*.

Chilhowee Lake, along the north shore of Fontana Lake at Pilkey Creek, and in Big Cove. King (1939b) also recorded this species at the old trout farming pond in Cataloochee; this site no longer exists.

Life History

Males call throughout the spring and summer from a variety of water bodies, including streams, beaver ponds, large temporary ponds, and even from the sewage treatment pond in Cades Cove. They have a distinctively deep, base voice: *jah harrrooom, jah harrrooom.* Some authors have likened it to "Jug o' Rum," giving rise to a colloquial name for this species. Because of the potentially long (1–2 years) larval period, some American Bullfrog tadpoles may be observed in suitable habitats throughout the year. Our survey crews heard American Bullfrogs calling from mid-May through August, with tadpoles commonly seen from early May to late September. At metamorphosis, the young frogs are already 31–59 mm (1.2–2.3 in) TL. Juveniles disperse widely in terrestrial habitats, but little is known about terrestrial habitat use by adults, especially during the winter. American

Bullfrogs have voracious appetites, and they eat anything that they can stuff into their mouths, including small mammals, birds, snakes, and other frogs, in addition to a varied diet of invertebrates. In areas where they occur naturally (the eastern United States and Canada), they likely have little effect on native vertebrates. In areas where they can be considered non-indigenous, their voracious appetite makes them a serious threat to native wildlife.

Abundance and Status

The American Bullfrog seems to be common in the lower regions of the Park where suitable habitat exists. Large numbers of tadpoles are observed each year in lower Abrams Creek near the Ranger Station, suggesting that viable reproduction is taking place. Elsewhere within its range, American Bullfrogs are numerous to the point of being a pest in areas within which they do not naturally occur.

Remarks

Adult male American Bullfrogs space themselves somewhat evenly along the shores of breeding sites. Their deep, booming call advertises their presence to females, as well as to warn rival males not to venture too close. If one male American Bullfrog invades the territory of another male, a fight will ensue.

Northern Green Frog
Rana clamitans

Etymology

Northern Green Frog adult, woodland pool behind Sugarlands Visitor Center.

Rana: Latin for frog; *clamitans:* from the Latin *clamito,* meaning to call loudly, or *clamator,* meaning bawler or shouter. Both names presumably are in reference to the vocal sounds produced by this frog, especially when trying to escape.

Northern Green Frog juvenile, Methodist Church Pond.

Northern Green Frog *(Rana clamitans).*

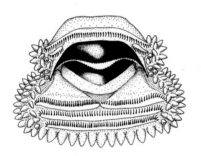

Northern Green Frog *(Rana clamitans).*

Identification

Adults. Northern Green Frogs are medium-sized, olive green to brownish frogs with a somewhat rugose dorsal skin surface and a dorso-lateral fold that extends from the rear of the eye halfway down the body. The dorsum, legs, and sides may be flecked with black spots. The tops of the thighs and legs do not possess black bars. On the side of the body, there is a dark, wavy, indistinct band that gives way ventrally to a mottled pattern. No light line is present on the upper lip. The belly is light but slightly mottled around its margins. The throat may be yellow (in males) to light gray or white. Adults measure 51–102 mm (2–4 in) TL.

Larvae. The tadpoles are large but not deep bodied, and are olive green with large dark spots. Throats are usually white, and bellies are a deep cream color with no iridescence. The tail is green and mottled with brown.

Eggs. The black and white eggs are laid in a surface film; each film contains from 1,000 to 5,000 eggs. The surface film may float free, or it may be attached to aquatic vegetation. Although somewhat similar to the surface film egg masses of American Bullfrogs, there are many fewer eggs in Northern Green Frog masses. Hatching occurs in a few days.

Similar Species. The Northern Green Frog most closely resembles the Pickerel and Northern Leopard Frogs in shape, size, and coloration. Pickerel Frogs and Northern Leopard Frogs have large spots that occur in two

well-defined parallel rows down the back between dorso-lateral folds; Northern Green Frogs are mostly un-spotted and have only partial dorso-lateral folds. The American Bullfrog lacks the dorso-lateral folds alto-gether. The tadpoles of these three species are difficult to differentiate and the species accounts and illustra-tions should be consulted for accu-rate identification. The tadpoles of *R. clamitans* are most similar to but darker than those of *R. pipiens.*

Northern Green Frog egg film, Gourley Pond in Cades Cove.

Taxonomic Comments. Two subspecies of the Green Frog currently are rec-ognized; *R. c. melanota* is the subspecies that occurs in the Great Smoky Mountains.

Distribution

Green Frogs occur from the island of Newfoundland south to northern Florida and west to the eastern margin of the Great Plains. In the Great Smokies, Green Frogs occur at lower elevations on the Tennessee side of the park, particularly in Cades Cove and in the Abrams Creek watershed. Green Frogs should be found on the North Carolina side of the Park, but there do not appear to be any recent records. Huheey and Stupka (1967) stated that they occur at elevations to 1,220 m (4,000 ft), but it is unclear whether they meant within the park or elsewhere within the range of the species. No Green Frogs were observed during surveys between 1998 and 2001 above ca. 550 m (1,800 ft).

Life History

Males call especially from both temporary and permanent pools and ponds and are not as common in flowing water as is *R. catesbeiana.* In the Great Smokies, males have been heard calling from late May to mid-July, although King (1939b) stated that Northern Green Frogs "makes its appearance in late April." Huheey and Stupka (1967) recorded breeding

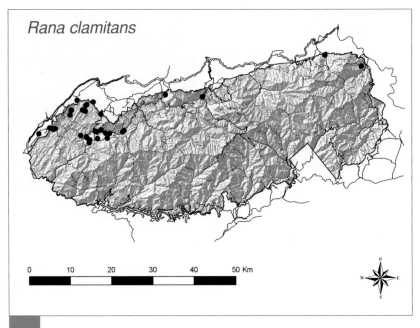

Distribution of the Northern Green Frog *(Rana clamitans)*.

activity as late as 27 July (1962) in Cades Cove. Large numbers of float-ing egg mats of this species were seen on 10 July 2000 at Gourley Pond. Tadpoles are commonly seen from late spring into the autumn. They may spend up to a year prior to metamorphosis, although most transform after several months. Metamorphosis occurs from July into the autumn.

It is not uncommon to find tadpoles of very different size classes within the same water body. Indeed, metamorphosis occurs at quite a range of sizes. For example, my wife and I observed very small metamorphic juve-niles (ca. 15 mm [0.6 in] SVL) at Meadow Branch on 5 August 2000, as well as large metamorphic juveniles (ca. 35 mm [1.4 in] SVL) at Methodist Church Pond on 20 September 2001. Both seemed to be in the same stage of development despite their size difference. The habitats also were very dif-ferent (shallow, drying, woodland pools versus a very richly organic farm pond, respectively), suggesting that resources varied substantially between the breeding sites. The presence of small tadpoles in late September at Methodist Church and Sugarlands Ponds indicates that some tadpoles over-winter in permanent water habitats within the Great Smokies.

Abundance and Status

The Northern Green Frog is commonly found at the lower elevations of the park, and is present in a variety of wetland habitats.

Remarks

Sometimes adults of this species are found far from the nearest breeding site. For example, we have observed *R. clamitans* inside the entrance to Gregory's Cave in Cades Cove on several occasions (Dodd, Griffey, and Corser 2001), at least 500 m (1,640 ft) from the nearest potential breeding site. Adults apparently wander considerable distances in terrestrial habitats, but little empirical information is available. In the northern portions of its range, *R. clamitans* leaves the vicinity of the breeding ponds in autumn to forage prior to overwintering. This allows the frog to build up its lipid reserves in order to survive the long winter. They overwinter in areas with flowing water in streams and seeps, where they find refuge under rocks, in tunnels, or hidden beneath root masses. In one instance, I discovered a large adult Northern Green Frog "sleeping" on the bottom of Abrams Creek among the rocks (8 August 2000). The animal was immobile until I lifted it from the water, at which point it seemed to suddenly awake and come alive. Searches failed to find additional adults on the river bottom.

Pickerel Frog
Rana palustris

Pickerel Frog adult, Mount Sterling Gap.

Etymology

Rana: Latin for frog; *palustris:* from the Latin *paluster,* meaning "of the marsh." The name literally means "a marsh frog."

Identification

Adults. This is a medium to large, olive-green to brownish frog with a light to cream-colored dorso-lateral fold that extends down the body from behind

Pickerel Frog transforming larva, woodland pool adjacent to Abrams Creek (September 2000).

Pickerel Frog *(Rana palustris)*.

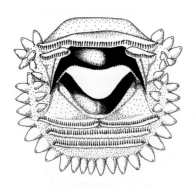

Pickerel Frog *(Rana palustris)*.

the eye to the groin. There are paired, large, squarish dorsal spots on the back and sides; these spots extend in two rows down the back between the dorso-lateral folds. The rear limbs have black bars giving a banded pattern. There is no white spot in the center of the tympanum, but a white line is present on the posterior part of the upper lip. The snout is pointed and the eyes are large. The belly and throat are white, but the underside of the hind limbs and groin may be yellow to orange. Males have paired vocal sacs. The largest known Pickerel Frog, 87 mm (3.4 in) SVL, was collected along Tater Branch in Cades Cove, but adults are mature by 44 mm (1.7 in) TL.

Larvae. The tadpole is large, deep, and full-bodied. The dorsal color is olive green grading to yellow on the sides. The belly is cream-colored with white blotches, whereas the dorsum is marked with fine black and yellow spots. The belly is iridescent and the viscera are visible. The tail is very dark with black blotches. When viewed from above, the body is very round or oval.

Eggs. The eggs are deposited in a firm, spherical cluster 38–100 mm (1.5–4 in) in diameter; each cluster contains 2,000–4,000 eggs. Egg masses are placed in deep water and are attached to debris and vegetation. Individual eggs are bicolored, brown above and yellow below.

Similar Species. The Northern Leopard Frog most closely resembles the Pickerel Frog in shape, size, and coloration. In Pickerel Frogs, the dorsal spots are squared rather than rounded, occur in two well-defined, parallel rows down the back between the dorso-lateral folds, and are paired rather than scattered about. These frogs also have distinct yellowish color on the underside of the thighs and groin, unlike the leopard frogs. The American Bullfrog lacks the dorso-lateral folds altogether, and the Northern Green Frog has only partial dorso-lateral folds and is unspotted. The tadpoles of these four species are difficult to differentiate. The throat of the Northern Leopard Frog tadpole is more extensive and translucent than that of the Pickerel Frog. Tadpoles of *R. palustris* usually contain a yellow wash on the sides of the body.

Taxonomic Comments. Although there is considerable geographic variation within this species, no subspecies are recognized.

Distribution

Pickerel frogs are found from the St. Lawrence River south to the South Carolina coast at the Savannah River, north of the Fall Line in the south Gulf Coast states, and in central North America from eastern Texas north to Wisconsin and Ontario. They are absent from the prairie regions of Illinois.

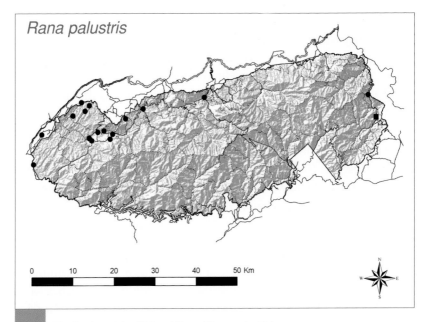

Distribution of the Pickerel Frog *(Rana palustris)*.

In the Great Smoky Mountains, Pickerel Frogs are found commonly on the Tennessee side of the park in the Little River, Fighting Creek, and Abrams Creek drainages, and in Cades Cove. King (1939b) reported Pickerel Frogs from the Caldwell Fork–Cataloochee Creek drainage on the North Carolina side. In an artificial pond along Cataloochee Creek, he stated that "the species is present by the thousands" (1939b:568); this pond no longer exists. Recent surveys could not find Pickerel Frogs at this site, now nothing more than a mucky wetland. However, we found an adult Pickerel Frog at Mount Sterling Gap (854 m [3,888 ft] on 6 June 1999. This location is approximately 5.4 km (3.2 mi) northwest of the former Cataloochee breeding pond, suggesting that Pickerel Frogs still occur in the area. Huheey and Stupka (1967) stated that the Pickerel Frog is widely distributed below 915 m (3,000 ft), but nearly all recent records are from areas below 610 m (2,000 ft).

Life History

The Pickerel Frog breeds at temporary ponds and woodland pools from early spring into summer, depending upon the availability of water. The voice is a "low-pitched grating croak with little carrying power" (Wright and Wright 1949). Males sometimes call from under the water. King (1939b) observed egg deposition on 21 March 1935 in Cades Cove, and Jeff Corser found egg masses at Gourley Pond on 29 March 2001. We have observed tadpoles at Gourley Pond from early May into mid-June. The larval period in other locations is 70 to 80 days. Metamorphosis occurs throughout the summer, depending upon time of egg deposition. I have seen metamorphs, which are 19–27 mm (0.75– 1.1 in) TL, on 29 June (1999) at Gourley Pond, and my wife, Marian, found recent metamorphs in a woodland pool near Abrams Creek on 10 September 2000; Huheey and Stupka (1967) collected metamorphs on 20 July 1962 in Cades Cove. Subadults are commonly observed in the autumn in low wetlands.

Abundance and Status

There are no empirical data on the abundance of this species within the Great Smokies. They were certainly more abundant in Cataloochee when the old trout pond was structurally sound. This species may be somewhat limited in its distribution by its preference for deeper, quiet waters in which to breed. Much more data are needed on the life history and habitat use by Pickerel Frogs within the Great Smoky Mountains.

Remarks

As with certain other frogs (Dodd, Griffey, and Corser 2001), Pickerel Frogs are encountered occasionally at the entrance to caves, where they find cool temperatures, high humidity, shelter, and food. Adults apparently wander considerable distances in terrestrial habitats, but little empirical information is available.

Pickerel Frogs also have a toxic and/or noxious secretion to which other amphibians are particularly susceptible. Pickerel Frogs are generally harmless to people, although some individuals and their curious pets may have an adverse reaction to the secretion. Never put your fingers into your eyes and always wash your hands after handling Pickerel Frogs.

Northern Leopard Frog
Rana pipiens

Etymology

Rana: Latin for frog; *pipiens:* from the Latin *pipiens,* meaning peeping. Apparently the first collector of this species heard Spring Peepers and thought that the leopard frogs that he had just collected were responsible for the loud peeps that he heard.

Northern Leopard Frog adult, near Logan, West Virginia. Photo by Richard Bartlett.

Identification

Adults. This is a medium to large, olive-green frog with a yellow to cream-colored dorso-lateral fold that extends the entire length of the body from the snout to the groin. There are scattered large, dorsal, unpaired spots on the back and sides; these spots may extend in two or three irregular rows down the back between the dorso-lateral folds. The rear limbs, especially, have black bars, creating a banded pattern. There is a white spot in the center of the tympanum, but the spot may be unclear or inconspicuous,

Northern Leopard Frog *(Rana pipiens)*.

and a white line is present on the upper lip. The snout is pointed and the eyes are large. The belly, throat, and underside of the hind limbs are white. Males have paired vocal sacs. Adults measure 51–102 mm (2–4 in) TL.

Larvae. The tadpole is large and deep bodied, dark brown dorsally, and covered with small gold spots. The belly is cream-colored with a bronze iridescence, the viscera tend to be visible through the skin, and the throat is translucent.

Northern Leopard Frog *(Rana pipiens)*.

Eggs. Eggs are deposited in a firm jellied mass, with each mass containing 3,500–6,500 eggs. The eggs are bicolored, white on the bottom and black on top, and tend to be packed close together. The egg mass is usually attached to grasses or submerged vegetation near the water's surface.

Similar Species. The Northern Leopard Frog most closely resembles the Pickerel Frog in shape, size, and coloration. In Pickerel Frogs, however, the dorsal spots are squared rather than rounded, occur in two well-defined, parallel rows down the back between the dorso-lateral folds, and are paired rather than scattered about. The American Bullfrog lacks the dorso-lateral folds altogether, and the Northern Green Frog has only partial dorso-lateral folds and is unspotted. The tadpoles of these three species are difficult to differentiate. Tadpoles of *R. clamitans* are similar to but darker than those of *R. pipiens;* the throat of the Northern Leopard Frog tadpole is more extensive and translucent than that of *R. palustris;* and the American Bullfrog tadpole has evenly spaced, very fine, dark spots on a lighter olive green background.

Taxonomic Comments. There is some confusion concerning the identity of the leopard frogs found historically in the Great Smokies. Standard field guides (e.g., Conant and Collins 1991; Redmond and Scott 1996) show range maps indicating that the Smokies' leopard frog should be the

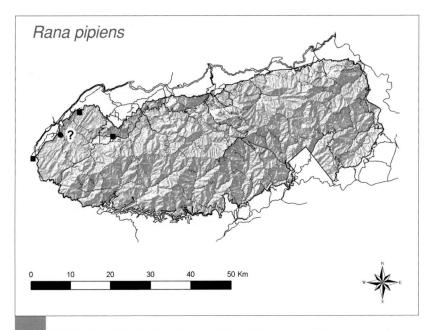

Distribution of the Northern Leopard Frog *(Rana pipiens)*. The squares show the position of historic locations: former town of Chilhowee (far left); Cane Creek (middle); Cades Cove (right). The "?" indicates an unverified sighting along Rabbit Creek Trail at Abrams Creek.

Southern Leopard Frog, *Rana sphenocephala.* However, I concur with King (1939b) in that the leopard frogs in the preserved collection of the Great Smoky Mountains National Park more closely resemble the Northern Leopard Frog rather than the Southern Leopard Frog. Until living frogs can be examined from the park, the systematic status of this species will remain unclear.

Distribution

The Northern Leopard Frog ranges from mainland Labrador (an introduced population) and Québec south to disjunct populations in West Virginia and Kentucky, west to Arizona and Nevada, and north to the MacKenzie District, Canada. If the Great Smokies leopard frog is indeed the Northern Leopard Frog, it probably represents an isolated population. As such, it would be the most southern location for this species in eastern North America.

As reported by Necker (1934) and King (1939b), leopard frogs from the Great Smoky Mountains were collected in Cades Cove (exact location not specified), near the former town of Chilhowee (274 m [900 ft]), and in the Cane Creek watershed (381 m [1,250 ft]). The small town of Chilhowee actually was located outside the park's boundary and was inundated when Chilhowee Lake was formed. A recent sighting (on 27 May 2000) near the Abrams Campground along lower Abrams Creek needs confirmation. This species might still occur along inlets on the northern shore of Chilhowee Lake or elsewhere at low elevations (lower than 550 m [1,800 ft]) in the Abrams–Cane–Beard Cane Creek drainages. However, surveys conducted from 1998 to 2001 by USGS personnel along Abrams Creek, Cane Creek, Chilhowee Lake, the north shore of Fontana Lake, and in Cades Cove did not result in any sightings, either of adults or tadpoles, or in any calling frogs that could be heard.

Life History

Because of the paucity of records, nothing is known concerning the life history of leopard frogs within Great Smoky Mountains National Park. All of the specimens reported thus far were recorded in winter or spring (4 May 1933 in Cades Cove; 13 February 1935 at Chilhowhee; 19 April 1935 on Cane Creek; 27 May 2000 at Abrams Creek) suggesting that movements occur early in the year, presumably to and from breeding sites. In other locations, the Northern Leopard Frog breeds from mid-March to April, and the larval period extends from 70 to 80 days. Permanent, quiet water ponds, sloughs, and lakes are the preferred breeding sites. At metamorphosis, the young frogs are about 19 to 27 mm (0.75 to 1.1 in) TL. The call is a drawn-out chuckle; one herpetologist described it as "a rattling snore, interspersed with assorted clucks and grunts, of one or more syllables" (Barbour 1971).

Abundance and Status

If this species still occurs within the boundary of Great Smoky Mountains National Park, it appears to be exceedingly scarce.

Remarks

When alarmed, these frogs give off a warning yelp as they leap into the water. Presumably, this serves to alert other frogs in the vicinity as to the

Smokies. It is likely that the numbers of egg masses produced from one year to the next varies considerably, especially with regard to hydroperiod, weather, and available food resources.

Remarks

Although opportunistic detritivores (that is, they eat all kinds of pond detritus, including plankton and algae), Wood Frog tadpoles occasionally feed on the jelly masses left by *Ambystoma maculatum* after hatching. On 12 April 2000, my wife and I observed hundreds of Wood Frog tadpoles eating old salamander jelly masses in roadside pools along the road to Tremont.

Occasionally ponds dry before Wood Frog tadpoles can complete metamorphosis. This could be a problem for Wood Frog tadpoles, which often occur in high densities, because growth rates decline as density increases. When hydroperiods are short, literally thousands of tadpoles may die from desiccation and freezing temperatures. In years when ponds do not fill in the winter or spring, a situation which occurred for many ponds in 1999, no reproduction takes place. From 1998 to 2001, survey crews observed stranded Wood Frog egg masses at a number of sites within the Great Smokies, including pools within the fields at Cataloochee, Gourley Pond, and at Gum Swamp.

Another threat to egg masses is fungal infections, such as have been observed on Wood Frog egg masses in the Cane Creek region; on 22 March 1999, fully 75 percent of the egg masses were found to be infected, resulting in high mortality. Jeff Corser (pers. comm.) also estimated 50 percent mortality from fungus on Wood Frog eggs found 14 February 2000 at Pinnacle Johns near the junction with Byrd Creek.

As noted by Huheey and Stupka (1967), large numbers of these frogs are killed on park roads as they migrate to and from breeding sites. Raccoons also can take a considerable toll during the breeding season. On 28 February 1940, Arthur Stupka "counted over 100 dead and injured Wood Frogs along the margin of a sinkhole pond near Laurel Creek" (Huheey and Stupka 1967), and Jeff Corser reported mass mortality at this same location on 12 February 2001. Dead Wood Frogs with tooth marks or that are partially eaten are sometimes found at other sites in the Smokies during the breeding season, including Gourley Pond and Gum Swamp (63 observed killed on 15 February 2001; J. Corser, pers. comm.).

Eastern Spadefoot Adult, male,
Gum Swamp.

Eastern Spadefoot
(*Scaphiopus holbrooki*).

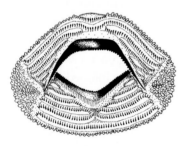

Eastern Spadefoot
(*Scaphiopus holbrooki*).

Eastern Spadefoot
Scaphiopus holbrooki

Etymology

Scaphiopus: from the Greek *skaphis,* meaning shovel or spade, and *pous,* meaning foot. The name is in reference to the spade on each hind foot, an adaptation for digging; *holbrooki:* in honor of South Carolina physician John Edwards Holbrook (1794–1871), considered the father of North American herpetology. Between 1836 and 1842, Holbrook published the first major review of the amphibians and reptiles of North America in a series of five separately issued parts (second edition, 1842; reprinted, 1976). Although commonly called a spadefoot toad, this species is in the family Pelobatidae, not the family Bufonidae, as are the true toads.

Identification

Adults. The Eastern Spadefoot is a medium-sized frog with a short, blunt snout and large, expressive eyes. The dorsum is brown and rugose, with light yellow, narrow lines running parallel down its back. The belly is white and unmarked. The rear feet of *Scaphiopus holbrooki* have obvious black spade-like projections that are used to dig backwards into the ground, giving this species its common name. This species has a very musty or peppery smell, to which some people are allergic. Adults measure 45–72 mm (1.8–2.8 in) TL.

Larvae. The tadpole is small and dark-colored, bronze to brown, with very small orange spots. The body is round when viewed from above, and the

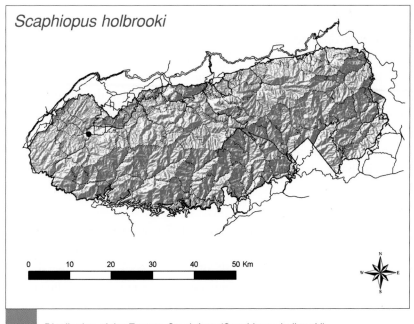

Distribution of the Eastern Spadefoot *(Scaphiopus holbrooki).*

eyes are close set. The head is wide relative to the body width, and the tail is short and rounded.

Eggs. The eggs are deposited in a loose, irregular cylinder or band about 25 to 75 mm (1 to 3 in) wide and 25 to 300 mm (1 to 11.8 in) long. About 200 eggs are oviposited per cylinder. The eggs are placed on stems or branches in the water.

Similar Species. This species cannot be confused with any other species in the Great Smoky Mountains. The dark-colored tadpoles are more rounded than those of the toads *(Bufo)* when viewed from above, and the snout is not pointed anteriorly when viewed from the side, as are the tadpoles of the Eastern Narrow-mouthed Toad *(Gastrophryne).*

Taxonomic Comments. Two subspecies of spadefoots have been described; the Eastern Spadefoot, *S. holbrooki holbrooki,* is the subspecies that occurs in the Great Smokies. Some biologists consider Hurter's Spadefoot, *S. h. hurterii,* of the south-central United States to be a separate species.

Distribution

The Eastern Spadefoot is found from Cape Cod south to the Florida Keys, and west to eastern Louisiana, Arkansas, Missouri, and southern Illinois. It is not common in the mountains of the southern Appalachians, where it is confined to lowland valleys. In the Great Smokies, the Eastern Spadefoot is known only from Gum Swamp in Cades Cove. Males were heard in chorus there on 12 July 1999, and a recently transformed juvenile was found on 20 September 1999 (Irwin, Dodd, and Griffey 1999). It seems unlikely that Eastern Spadefoots occur elsewhere within the park.

Life History

In most areas, Eastern Spadefoots are explosive breeders, and breeding only occurs after the heaviest of rains, as it did from 10 to 12 July 1999. The call has been described as a "wank wank" or a "wonk wonk," but neither description does justice to it. Males, when handled, have a pitious-sounding release call. The larval period is short (as few as 14 days), and juveniles (measuring 8.5–12 mm [0.3–0.5 in] TL at metamorphosis) and adults disperse rapidly from the breeding ponds. As larvae, the tadpoles form large schools that aggregate and swim together. Presumably, these schools function much the same way as fish schools do, i.e., they make an individual less vulnerable to a predator. Schooling also may help churn up detritus on which the tadpoles feed. This is also one of the few North American species whose tadpoles are cannibalistic. Another interesting facet of the larval stage is that the tadpoles within a school are related to one another, each resulting from the same egg clutch. When schools are mixed, they tend to sort out with their siblings and avoid non-related tadpoles. The adults live a somewhat fossorial existence, emerging only to the mouths of their burrows to feed; they are rarely seen away from breeding ponds.

Abundance and Status

Eastern Spadefoots are extremely rare within the Great Smokies. However, there are no known threats to them throughout their range, although habitat destruction and alteration must surely take a toll. They are far more commonly seen in the southern portions of their range than in the north, where breeding may take place only once every few years depending upon rainfall patterns.

Duellman, W. E., and S. S. Sweet. 1999. Distribution patterns of amphibians in the Nearctic Region of North America. *In* W. E. Duellman, ed. Patterns of Distribution of Amphibians. A Global Perspective, pp. 31–109. Johns Hopkins Univ. Press, Baltimore, MD.

Dunn, E. R. 1926. The Salamanders of the Family Plethodontidae. Smith College Publications, Northampton, MA. 441 pp. Reissued in 1972 by the Society for the Study of Amphibians and Reptiles.

Frisbie, M. P., and R. L. Wyman. 1991. The effects of soil pH on sodium balance in the red-backed salamander, *Plethodon cinereus,* and three other terrestrial salamanders. Physiological Zoology 64:1050–1068.

Frisbie, M. P., and R. L. Wyman. 1992. The effect of environmental pH on sodium balance in the red-spotted newt, *Notophthalmus viridescens.* Archives of Environmental Contamination and Toxicology 23:64–68.

Frost, D. R. 2000. Species, descriptive efficiency, and progress in systematics. *In* R. C. Bruce, R. G. Jaeger, and L. D. Houck, eds. The Biology of Plethodontid Salamanders, pp. 7–29. Kluwer Academic, New York, NY.

Frost, D. R. and D. M. Hillis. 1990. Species in concept and practice: herpetological applications. Herpetologica 46:87–104.

Gordon, R. E. 1952. A contribution to the life history and ecology of the plethodontid salamander *Aneides aeneus* (Cope and Packard). The American Midland Naturalist 47:666–701.

Hairston, N. G., Sr. 1983. Growth, survival, and reproduction of *Plethodon jordani:* trade-offs between selective pressures. Copeia 1983:1024–1035.

Hairston, N. G., Sr. 1987. Community Ecology and Salamander Guilds. Cambridge Univ. Press, Cambridge, United Kingdom. 230 pp.

Hairston, N. G., Sr. 1993. On the validity of the name *teyahalee* as applied to a member of the *Plethodon glutinosus* complex (Caudata: Plethodontidae): a new name. Brimleyana 18:65–69.

Hairston, N. G., Sr., and R. H. Wiley. 1993. No decline in salamander (Amphibia: Caudata) populations: a twenty-year study in the Southern Appalachians. Brimleyana 18:59–64.

Hairston, N. G., Sr., R. H. Wiley, and C. K. Smith. 1992. The dynamics of two hybrid zones in Appalachian salamanders of the genus *Plethodon.* Evolution 46:930–938.

Hassler, W. G. 1929. Salamanders of the Great Smokies. Natural History 29(1): 95–100.

Hedges, S. B. 1986. An electrophoretic analysis of Holarctic hylid frog evolution. Systematic Zoology 35:1–21.

Hensel, J. L., Jr., and E. D. Brodie, Jr. 1976. An experimental study of aposematic coloration in the salamander *Plethodon jordani.* Copeia 1976:59–65.

Highton, R. 1989. Biochemical evolution in the slimy salamanders of the *Plethodon glutinosus* complex in the eastern United States. Part 1. Geographic protein variation. Univ. of Illinois Biological Monographs 57:1–78.

Highton, R. 1997. Geographic protein variation and speciation in the *Plethodon dorsalis* complex. Herpetologica 53:345–356.

Highton, R., and R. B. Peabody. 2000. Geographic protein variation and speciation in salamanders of the *Plethodon jordani* and *Plethodon glutinosus* complexes in the Southern Appalachian Mountains with the description of four new species. *In* R. C. Bruce, R. G. Jaeger, and L. D. Houck, eds. The Biology of Plethodontid Salamanders, pp. 31–93. Kluwer Academic, New York, NY.

Hofrichter, R., ed. 2000. Amphibians. The World of Frogs, Toads, Salamanders and Newts. Firefly Books, Buffalo, NY. 264 pp.

Holman, J. A. 1995. Pleistocene Amphibians and Reptiles in North America. Oxford Univ. Press, New York. 243 pp.

Houk, R. 1993. Great Smoky Mountains National Park. A Natural History Guide. Houghton Mifflin Company, NY.

Howard, R. R., and E. D. Brodie, Jr. 1973. A Batesian mimetic complex in salamanders: responses to avian predators. Herpetologica 29:33–41.

Howe, T. D., and S. P. Bratton. 1976. Winter rooting activity of the European wild boar in the Great Smoky Mountains National Park. Castanea 41:256–264.

Huheey, J. E. 1964. Use of burrows by the black-bellied salamander. Journal of the Ohio Herpetological Society 4:105.

Huheey, J. E. 1966. The desmognathine salamanders of the Great Smoky Mountains National Park. Journal of the Ohio Herpetological Society 5:63–72.

Huheey, J. E., and A. Stupka. 1967. Amphibians and Reptiles of Great Smoky Mountains National Park. Univ. of Tennessee Press, Knoxville. 98 pp.

Hunsinger, T. W. 2001. The writings of Sherman Bishop: Part II. Conservation. Herpetological Review 32:241–244.

Hyde, E. J. 2000. Assessing the diversity and habitat associations of salamanders in Great Smoky Mountains National Park. M.S. thesis, North Carolina State Univ., Raleigh. 103 pp.

Hyde, E. J., and T. R. Simons. 2001. Sampling plethodontid salamanders: sources of variability. Journal of Wildlife Management 65:624–632.

Irwin, K. J., C. K. Dodd, Jr., and M. L. Griffey. 1999. Geographic distribution: *Scaphiopus holbrooki* (Eastern spadefoot toad). Herpetological Review 30:232.

Jacobs, J. F. 1987. A preliminary investigation of geographic genetic variation and systematics of the two-lined salamander, *Eurycea bislineata* (Green). Herpetologica 43:423–446.

Kemp, S. 1993. Trees of the Smokies. Great Smoky Mountains Natural History Association, Gatlinburg, Tennessee. 125 pp.

King, W. 1936. A new salamander *(Desmognathus)* from the Southern Appalachians. Herpetologica 1:57–60.

King, W. 1939a. A herpetological survey of the Great Smoky Mountains National Park. Ph.D. diss., Univ. of Cincinnati, Cincinnati, OH. 77 pp.

King, W. 1939b. A survey of the herpetology of Great Smoky Mountains National Park. The American Midland Naturalist 21:531–582.

King, W. 1944. Additions to the list of amphibians and reptiles of Great Smoky Mountains National Park. Copeia 1944:255.

Knapp, R. A., and K. R. Matthews. 2000. Non-native fish introductions and the decline of the mountain yellow-legged frog from within protected areas. Conservation Biology 14:428–438.

Koenings, C. A., C. K. Smith, E. A. Domingue, and J. W. Petranka. 2000. Natural History Notes: *Desmognathus imitator* (Imitator Salamander). Reproduction. Herpetological Review 31:38–39.

Kozak, K. H., and R. R. Montanucci. 2001. Genetic variation across a contact zone between montane and lowland forms of the Two-lined Salamander *(Eurycea bislineata)* species complex: a test of species limits. Copeia 2001:25–34.

Kuchen, D. J., J. S. Davis, J. W. Petranka, and C. K. Smith. 1994. Anakeesta stream acidification and metal contamination: effects on a salamander community. Journal of Environmental Quality 23:1311–1317.

Labanick, G. M., and R. A. Brandon. 1981. An experimental study of Batesian mimicry between the salamanders *Plethodon jordani* and *Desmognathus ochrophaeus*. Journal of Herpetology 15:275–281.

Lydic, J. 1999. Populations of salamanders within an old and secondary growth mesic cove forest with reference to coarse woody debris. M.S. thesis, Edinboro Univ. of Pennsylvania, Edinboro. 66 pp.

Martof, B. S. 1956. Three new subspecies of *Leurognathus marmorata* from the southern Appalachian Mountains. Occasional Papers of the Museum of Zoology, Univ. of Michigan. (575):1–13.

Martof, B. S., W. M. Palmer, J. R. Bailey, and J. R. Harrison III. 1980. Amphibians and Reptiles of the Carolinas and Virginia. Univ. of North Carolina Press, Chapel Hill. 264 pp.

Mathews, R. C., Jr. 1984. Distributional ecology of stream-dwelling salamanders in the Great Smoky Mountains National Park. M.S. thesis, Univ. of Tennessee, Knoxville. 186 pp.

Mathews, R. C., Jr., and E. L. Morgan. 1982. Toxicity of Anakeesta Formation leachates to shovel-nosed salamander, Great Smoky Mountains National Park. Journal of Environmental Quality 11:101–106.

Mathews, R. C., Jr., and A. C. Echternacht. 1984. Herpetofauna of the spruce-fir ecosystem in the Southern Appalachian Mountain regions, with emphasis on the Great Smoky Mountains National Park. *In* P. S. White, ed. The Southern Appalachian Spruce-Fir Ecosystem, pp. 155–167. National Park Service Research/Resource Management Report SER-71.

Matthews, K. R., K. L. Pope, H. K. Preisler, and R. A. Knapp. 2001. Effects of nonnative trout on Pacific treefrogs *(Hyla regilla)* in the Sierra Nevada. Copeia 2001:1130–1137.

McClure, G. V. 1931. The Great Smoky Mountains with preliminary notes on the salamanders of Mt. LeConte and LeConte Creek. Zoologica 11(6):53–76.

Means, D. B. 2000. Southeastern U.S. coastal plain habitats of the Plethodontidae. The importance of relief, ravines, and seepage. *In* R. C. Bruce, R. G. Jaeger, and L. D. Houck, eds. The Biology of Plethodontid Salamanders, pp. 287–302. Kluwer Academic, New York, NY.

Merchant, H. 1972. Estimated population size and home range of the salamanders *Plethodon jordani* and *Plethodon glutinosus.* Journal of the Washington Academy of Science 62:248–257.

Merkle, D. A., and D. A. Kovacic. 1974. A new record for *Necturus maculosus* in the Great Smoky Mountains. Journal of the Tennessee Academy of Science 49:142.

Moore, H. L. 1988. A Roadside Guide to the Geology of the Great Smoky Mountains National Park. Univ. of Tennessee Press, Knoxville.

National Park Service. 1984. At Home in the Smokies. A History Handbook for Great Smoky Mountains National Park, North Carolina and Tennessee. Division of Publications, National Park Service, Washington, DC. 159 pp.

National Park Service. 1999. Environmental Assessment. Brook Trout Restoration. Great Smoky Mountains National Park. Gatlinburg, Tennessee.

Necker, W. L. 1934. Contribution to the herpetology of the Smoky Mountains of Tennessee. Bulletin of the Chicago Academy of Sciences 5:1–4.

Nickerson, M. A., K. L. Krysko, and R. D. Owen. 2002. Ecological status of the Hellbender *(Cryptobranchus alleganiensis)* and the Mudpuppy *(Necturus maculosus)* salamanders in the Great Smoky Mountains National Park. Journal of the North Carolina Academy of Science 118:27–34.

Nishikawa, K. C. 1990. Intraspecific spatial relationships of two species of terrestrial salamanders. Copeia 1990:418–426.

Oliver, D. 1989. Hazel Creek from Then till Now. 133 pp. Available from the Great Smoky Mountains Natural History Association, Gatlinburg, Tennessee.

Organ, J. A. 1961. Life history of the pygmy salamander, *Desmognathus wrighti*, in Virginia. The American Midland Naturalist 66:384–390.

Petranka, J. W. 1998. Salamanders of the United States and Canada. Smithsonian Institution Press, Washington, DC. 587 pp.

Petranka, J. W., and S. S. Murray. 2001. Effectiveness of removal sampling for determining salamander density and biomass: a case study in an Appalachian streamside community. Journal of Herpetology 35:36–44.

Petranka, J. W., M. E. Eldridge, and K. E. Haley. 1993. Effects of timber harvesting on Southern Appalachian salamanders. Conservation Biology 7:363–370.

Petranka, J. W., M. P. Brannon, M. E. Hopey, and C. K. Smith. 1994. Effects of timber harvesting on low elevation populations of southern Appalachian salamanders. Forest Ecology and Management 67:135–147.

Powders, V. N., and W. L. Tietjen. 1974. The comparative food habits of sympatric and allopatric salamanders, *Plethodon glutinosus* and *Plethodon jordani* in eastern Tennessee and adjacent areas. Herpetologica 30:167–175.

Ptacek, M. B., H. C. Gerhardt, and R. D. Sage. 1994. Speciation by polyploidy in treefrogs: multiple origins of the tetraploid, *Hyla versicolor*. Evolution 48: 898–908.

Reagan, N. L., and P. A. Verrell. 1991. The evolution of plethodontid salamanders: did terrestrial mating facilitate lunglessness? American Naturalist 138:1307–1313.

Redmond, W. H., and A. F. Scott. 1996. Atlas of Amphibians of Tennessee. The Center for Field Biology, Miscellaneous Publication 12, 94 pp.

Resetarits, W. J., Jr. 1997. Differences in an ensemble of streamside salamanders (Plethodontidae) above and below a barrier to brook trout. Amphibia—Reptilia 18:15–25.

Ruben, J. A., and A. J. Boucot. 1989. The origin of the lungless salamanders (Amphibia: Plethodontidae). American Naturalist 134:161–169.

Ruben, J. A., N. L. Reagan, P. A. Verrell, and A. J. Boucot. 1993. Plethodontid salamander origins: a response to Beachy and Bruce. American Naturalist 142:1038–1051.

Ryan, T. J. 1997. Larva of *Eurycea junaluska* (Amphibia: Caudata: Plethodontidae), with comments on distribution. Copeia 1997:210–215.

Ryan, T. J. 1998. Larval life history and abundance of a rare salamander, *Eurycea junaluska*. Journal of Herpetology 32:10–17.

Schmidt, R. G., and W. S. Hooks. 1994. Whistle Over the Mountain. Timber, Track & Trails in the Tennessee Smokies. Graphicom Press, Yellow Springs, OH. 170 pp.

Semlitsch, R. D. 1998. Biological delineation of terrestrial buffer zones for pond-breeding salamanders. Conservation Biology 12:1113–1119.

Semlitsch, R. D. 2000. Principles for management of aquatic-breeding amphibians. Journal of Wildlife Management 64:615–631

Sever, D. M. 1983. Observations on the distribution and reproduction of the salamander *Eurycea junaluska* in Tennessee. Journal of the Tennessee Academy of Science 58:48–50.

Smith, C. K., and J. W. Petranka. 2000. Monitoring terrestrial salamanders: repeatability and validity of area-constrained cover object searches. Journal of Herpetology 34:547–557.

Smith, G. F., and N. S. Nicholas. 1998. Patterns of overstory composition in the fir and fir-spruce forests of the Great Smoky Mountains after balsam woolly adelgid infestation. The American Midland Naturalist 139:340–352.

Starnes, S. M., C. A. Kennedy, and J. W. Petranka. 2000. Sensitivity of embryos of Southern Appalachian amphibians to ambient solar UV-radiation. Conservation Biology 14:277–282.

Tilley, S. G. 1981. A new species of *Desmognathus* (Amphibia: Caudata: Plethodontidae) from the Southern Appalachian mountains. Occasional Paper of the University of Michigan Museum of Zoology (695):1–23.

Tilley, S. G. 1988. Hybridization between two species of *Desmognathus* (Amphibia: Caudata: Plethodontidae) in the Great Smoky Mountains. Herpetological Monographs 2:27–39.

Tilley, S. G. 2000. The systematics of *Desmognathus imitator*. *In* R. C. Bruce, R. G. Jaeger, and L. D. Houck, eds. The Biology of Plethodontid Salamanders, pp. 121–147. Kluwer Academic, New York, NY.

Tilley, S. G., and J. R. Harrison. 1969. Notes on the distribution of the pygmy salamander, *Desmognathus wrighti* King. Herpetologica 25:178–180.

Tilley, S. G., and M. J. Mahoney. 1996. Patterns of genetic differentiation in salamanders of the *Desmognathus ochrophaeus* complex (Amphibia: Plethodontidae). Herpetological Monographs 10:1–42.

Tilley, S. G., and J. E. Huheey. 2001. Reptiles & Amphibians of the Smokies. Great Smoky Mountains Natural History Association, Gatlinburg, Tennessee. 143 pp.

Tilley, S. G., R. B. Merritt, B. Wu, and R. Highton. 1978. Genetic differentiation in salamanders of the *Desmognathus ochrophaeus* complex (Plethodontidae). Evolution 32:93–115.

Titus, T., and A. Larson. 1996. Molecular phylogenetics of desmognathine salamanders (Caudata: Plethodontidae): a reevaluation of evolution in ecology, life history, and morphology. Systematic Biology 45:451–472.

Waldron, J. L., C. K. Dodd, Jr., and J. D. Corser. 2003. Leaf litterbags: factors affecting capture of stream-dwelling salamanders. Applied Herpetology, 1:23–36.

Weals, V. 1993. Last Train to Elkmont. Olden Press, Knoxville, TN. 150 pp.

Weller, W. H. 1930a. A new salamander from the Great Smoky Mountain National Park. Proceedings of the Junior Society of Natural History, Cincinnati 1(7):3–4.

Weller, W. H. 1930b. Notes on *Aneides,* Cope and Packard. Proceedings of the Junior Society of Natural History, Cincinnati 1(1):2–3.

Weller, W. H. 1931. A preliminary list of the salamanders of the Great Smoky Mountains of North Carolina and Tennessee. Proceedings of the Junior Society of Natural History, Cincinnati 2(1):21–32.

Wood, J. T. 1946. Measurements of a giant *Pseudotriton montanus montanus* larva from Great Smoky Mountains National Park. Copeia 1946:168.

Wood, J. T., and F. E. Wood. 1955. Notes on the nests and nesting of the Carolina mountain dusky salamander in Tennessee and Virginia. Journal of the Tennessee Academy of Science 30:36–39.

Woods, F. W., and R. E. Shanks. 1959. Natural replacement of chestnut by other species in the Great Smoky Mountains National Park. Ecology 40:349–361.

Wright, A. H., and A. A. Wright. 1949. Handbook of Frogs and Toads of the United States and Canada. Third edition. Cornell Univ. Press, Ithaca, NY. 640 pp. Reprinted in 1995.

Wyman, R. L., and D. S. Hawksley-Lescault. 1987. Soil acidity affects distribution, behavior, and physiology of the salamander *Plethodon cinereus.* Ecology 68:1819–1827.

References

Index

The Amphibians of Great Smoky Mountains National Park was designed and typeset on a Macintosh computer system using QuarkXPress software. The body text is set in 10.5/14 Adobe Garamond and display type is set in Letter Gothic. This book was designed and typeset by Cheryl Carrington and manufactured by Four Colour Imports.